CW01653102

T207 Engineering: mechanics, materials, design

BLOCK 4

RESOURCES: FLUIDS AND ENERGY

PARTS 1, 2 AND 3

The Open University

Part 1
Power to the people

CONTENTS

continued ...

1 THE DEMAND FOR ELECTRICITY

1.1 Supply and demand

Simply to exist, any human being has a need for the basic resources of oxygen, food and water. Similarly, an industrialized society needs some basic 'life-giving' supplies. Among these vital resources are water and electricity – most buildings are hooked up to regional, if not national, networks that supply them. Electricity has become a versatile medium through which energy can be transported from one location to another. For instance, the chemical energy of a fuel can be converted to electrical energy in one location, transferred by means of electrical currents in a network of cables to some other location, and there used to generate mechanical energy, thermal energy (or heat) or otherwise do useful work. It is a familiar commodity in the developed world.

This block is concerned with some engineering issues associated with both water and electricity. In particular, we will be looking into an ingenious combination of them that solves one of the biggest problems faced by a supplier of any resource: how to match the supply to the demand.

National and regional stocks of water are fairly easily held in reservoirs that are based on natural lakes, or else use is made of artificial lakes created by means of dams. More locally, water towers and storage tanks simply hold water ready for use. In the UK the various reservoirs are filled with rain water that drains from the land over time. In terms of demand, the worst that can happen is that in times of drought, stocks run out. Stocks of electricity pose a greater challenge. There are no *simple* natural electrical equivalents of rain and reservoirs – all our electricity has to be made more or less on demand.

1.2 Variations in demand

Over a 24-hour period, there are times when domestic electricity usage is relatively low and times when it is much higher. Exercise 1.1 will get us thinking about why.

Exercise 1.1

(a) Table 1.1 shows a list of electrical appliances that might be used in a typical household. Complete the columns on energy consumption based on your own household.

(b) Describe qualitatively how you would expect the national *domestic* demand for electrical power to vary over 24 hours.

(c) Suggest how the demand for electrical power from the *manufacturing and transport industries* might vary over a 24-hour period.

Table 1.1 Possible electrical appliance usage in a typical household

Appliances	Typical power rating / kW	Energy consumed / kW h		
		Daytime	Evening	Night-time
TV, video, DVD, hi-fi, PC	0.2–0.5			
Vacuum cleaner	0.3			
DIY power tools	0.2–1.0			
Kitchen gadgets	0.2–3.0			
Fridge	0.3			
Freezer	0.6			
Washing machine	1.0–3.0			
Hot water (if electric only)	3.0			
Space heating (if electric only)	3.0			

Figure 1.1 shows a typical UK electrical power demand curve over a 24-hour period during the summer. Notice the lower night-time level and the higher daytime level. During the night, when most of the population is asleep, there is much less demand for electricity.

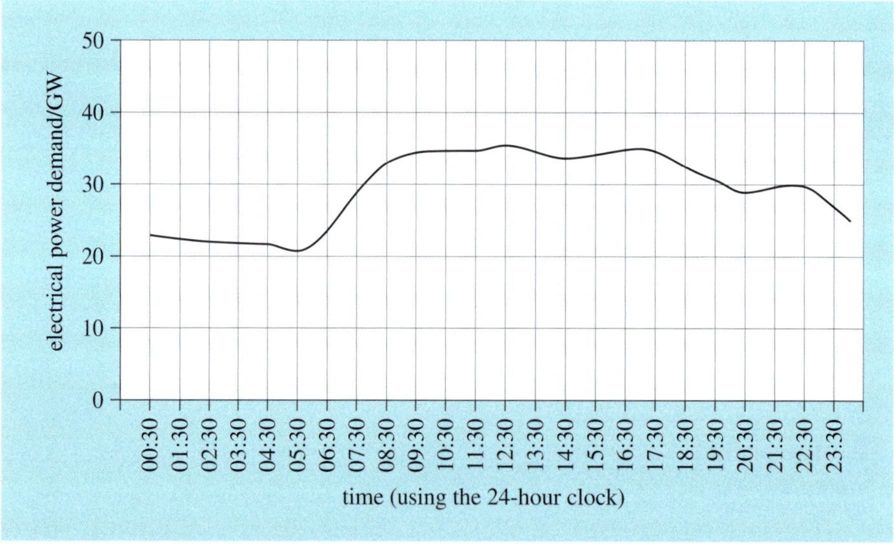

Figure 1.1 Typical instantaneous electrical power demand for the UK grid during a summer day

The electricity supply needs to be carefully planned to match demand. The consequence of not doing so would be periods when the mains supply voltage falls below acceptable levels owing to excessive demand – rather like the way that water pressure falls in a system that has too many taps open at one time. At times of low demand a huge amount of generating machinery is needlessly idling (that is, producing no electrical power).

Most of the UK's electricity is generated by fossil-fuel power stations. Planning and controlling their output is the chief strategy for coping with the diurnal variation (i.e. that during a day) seen in Figure 1.1. The capacity of the overall power system is reduced at night and increased during the day in anticipation of the regular predictable pattern, as described in **▼Capacity and output▲**.

▼Capacity and output▲

If you've ever taken part in a team activity like pulling a tug of war, pushing a broken-down car, rowing a boat or lifting a large object, you'll appreciate that you can appear to be contributing without actually delivering your fair share.

Here is another situation where you might have recognized that you were not contributing to the full. On a steady downhill section of cycling you can pedal very lightly, barely applying any pressure to the pedals, yet substantial speed is maintained.

In all these circumstances, you may have the capacity to provide substantially more output than you actually are doing. Power generating units hooked up together are similarly able to offer capacity without necessarily adding to the net power output. The actual output of electrical power depends on the mechanical power input. For a steam-driven turbine coupled to a generator, the output of electrical power is regulated through the supply of steam to the turbine.

However there is another problem. In addition to the variation during each day and from day to day, at certain times there are often sudden peaks in the domestic demand for electricity. These peaks are associated with a very large number of households, all at the same time, switching on a power-hungry electrical device; for instance they might all suddenly choose to boil water for making hot drinks. Another contribution to sudden peaks of demand occurs when millions of consumers all draw water around the same time. This causes pumping stations to switch on to restore the head of water in the distribution network. It ought not to be a problem – surely the preparation of hot drinks and the flushings of toilets and bathing are spread over fairly long intervals around and between meal times? Alas, as a population we do not act quite so randomly. Major synchronizing factors need to be considered. One of these, for example, is related to television broadcast schedules which in the UK can release several million people in search of a cup of tea or a bathroom within seconds of a popular television programme ending.

The sharp increases in electricity usage which occur during commercial breaks and at the ends of programmes are often referred to as *TV pickups*. In some ways this term is a little misleading – TV pickups are not caused by millions of television sets being turned on but by huge quantities of water being boiled or pumped at the end of a particular broadcast event. The highest-ever TV pickup in the UK during the twentieth century was 2800 MW on 4 July 1990, at the end of the football World Cup semi-final between England and West Germany that ended in a dramatic penalty shoot-out. Figure 1.2 shows a graph of UK electricity usage during the evening of the game. The notes on the graph explain the extra electricity being used.

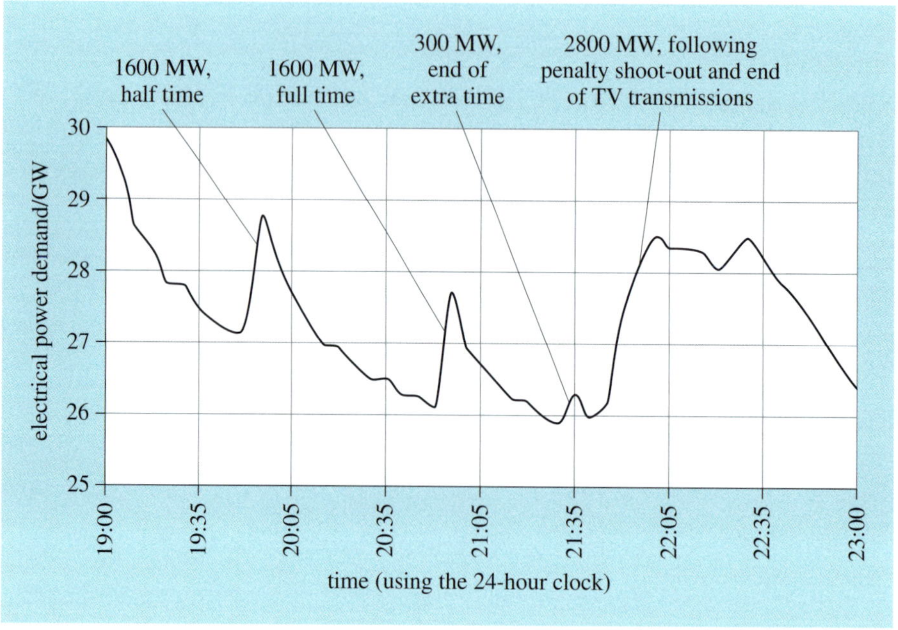

Figure 1.2 Instantaneous electrical power demand for the UK grid during an international football match in 1990. The match was the England v. West Germany World Cup semi-final, kick-off 19:00. At this time in the evening consumption is usually declining, as can be seen in Figure 1.1

Exercise 1.2

To boil 1 litre of tap water takes approximately 400 kJ of energy (I've supposed we need to heat it from about 5 °C).

(a) Calculate the average power consumed when 1 litre of water is boiled in 200 seconds.

(b) Following the 1990 World Cup semi-final, the TV pickup was 2800 MW. Let's assume this was entirely due to hot drinks being made. Estimate how many litres of water were boiled at the end of the match.

(c) Assuming that for each hot drink about 0.2 litres of water is boiled, estimate how many were made at the end of the match. (The UK population is about 60 million.)

There was an even greater increase in demand for electricity experienced in the UK on 11 August 1999 following the solar eclipse when within minutes demand dropped, and then surged by about 3000 MW. About 1000 MW of the increase was TV pickup – the TV audience making hot drinks and flushing toilets etc. as life returned to normal. The rest was due to the big 'switch on' as people returned to their normal activities at home and at work. Table 1.2 charts the variation in demand.

Table 1.2 Electricity demand during the morning of the solar eclipse on 11 August 1999

Time	10:30	10:40	10:50	11:00	11:10	11:15	11:20	11:30	11:40	11:50	12:00	12:10
Demand / MW	35 400	35 350	35 300	35 000	33 800	33 150	33 500	35 200	36 000	36 050	35 900	35 800

SAQ 1.1 (Outcomes 1.1 and 1.2)

Table 1.2 details the electricity usage during the morning of the solar eclipse on Wednesday 11 August 1999. Figure 1.3 shows how demand had varied over the same period *the day before*.

(a) Plot the data from Table 1.2 onto Figure 1.3 and label it appropriately.

(b) Suggest why, prior to the 3000 MW increase, the graph also shows a sudden decrease in electricity usage.

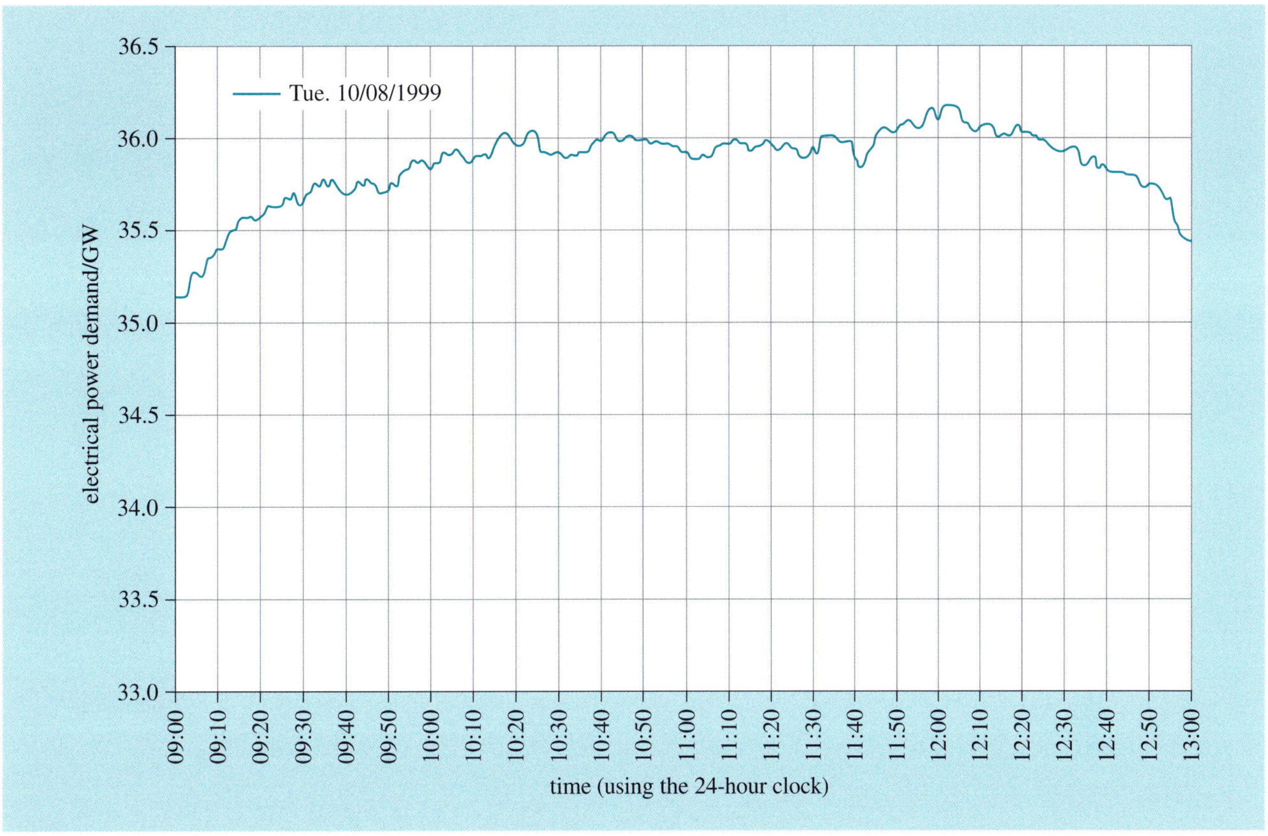

Figure 1.3 Instantaneous electrical power demand for the UK grid on the morning of *the day before* the solar eclipse on 11 August 1999

One of the UK's larger power stations, the Didcot 'A' coal/gas-fired power station, has a capacity of 2000 MW. You might think therefore that to cope with sudden increases in demand for electricity, all the suppliers need to do is have a couple of such power stations ready to be switched on when required. The problem is that it takes half an hour or so to bring a fossil-fuel power station up to full power. Nuclear power stations take longer. While this is adequate for coping with the predictable, daily variation in demand, it is far too slow when it comes to the fast, synchronized surges in demand such as TV pickups or unexpected failures in supply or in the grid. What's needed is something that can go from nothing to full power within seconds, and then keep up supplies until the other power stations catch up, or else shut down again quickly if the surge in demand is short-lived.

1.3 Aim of Part 1

The overall aim of this block is to illustrate the nature of problem solving in connection with the supply of electrical power generated by water-driven turbines. You will encounter more than one response to the challenge of supplying electricity rapidly at times when the demand increases suddenly.

Part 1 concerns the overall context. To get to grips with the whole problem it's important to have a clear appreciation of energy and the way electricity is generated, so we will look at these 'background' topics next in Sections 2 and 3. Then we can discuss ways of usefully holding in reserve sufficient energy to supply around a gigawatt of electrical power to the UK national electrical-power grid for several minutes, at a few seconds' notice. The approach outlined in Part 1 of Block 1 will be used to do this in Section 4.

In later parts of Block 4 the turbines and bearings that are needed to recover energy from practical, large-scale storage will be investigated.

2 ENERGY

Electricity in the present context is a means of transporting energy. So, when we discussed the demand for electricity in the previous section, what we were really talking about was a demand for energy. In fact, it was a demand specifically for electrical energy, delivered at a particular rate (i.e. a particular amount of electrical power). But what exactly is energy?

Energy was discussed in Block 3 Part 2, Section 3.3, from the point of view of mechanics and dynamics. The discussion here is broader.

2.1 What is energy?

Energy is the most precious commodity we have. Without it there could be no light, no warmth, no movement – no life. Everything we do involves energy. Anything that happens – from the flapping of a butterfly's wings to the eruption of a volcano – involves a change or transfer of energy (Figure 1.4). When we switch on a light, boil a kettle, drive a car, or go for a walk, we are using energy.

Figure 1.4 Energy and events

Energy is the capacity for making something happen or, in effect, for doing work. Based on your study of Block 3 you should know that for the purposes of dynamics energy is given by algebraic combinations such as:

$$\tfrac{1}{2}mv^2, \tfrac{1}{2}I\omega^2, I\alpha\theta \text{ and } mgh$$

and that it is properly measured in joules (or, on the atomic scale, in electron volts – Block 1 Part 2). But we need now to go beyond mechanistic formulas and conventional systems of units to be able to recognize and use the diversity of energy.

Most of the energy consumed on the Earth derives from the Sun. Every second, the Sun radiates about 3×10^{26} J of energy, of which about 10^{17} J reach the Earth. Some of this energy is transferred to plant and animal life where it is effectively stored; some of this energy is locked up in fossil fuels. Fossil fuels are so called because they are formed over millions of years from the fossils, or remains, of animals and plants. The Sun's energy is also responsible for creating wind in the Earth's atmosphere and for driving the water cycle.

Our bodies obtain energy from the food (plant and animal matter) that we eat. Meanwhile the energy we need for other tasks, such as heating and transport, can be obtained from the Sun directly through solar panels and photovoltaic cells, or indirectly by burning fossil fuels, or by extracting power from the wind or moving water to generate electricity. In the Sun, the energy was originally released through a series of nuclear reactions. I've been careful not to attribute the source of all our energy to the Sun. Mankind has learned how to control the release of energy from other nuclear reactions on the Earth, and these are another primary source of energy for our terrestrial life. And that's not all – tidal energy derives from the motion of the Moon around the Earth.

2.2 Forms of energy

Energy exists in different forms that are variously described or classified as mechanical, electrical, chemical, nuclear or thermal. Let's review what is meant by each of these terms.

2.2.1 Mechanical energy

Normally, if we say that something is mechanical, we are implying that it is either operated by a machine or it behaves in a machine-like manner, or else it incorporates mechanisms. Mechanical energy is the energy associated with movement or the potential for movement of such machinery or indeed any macroscopic object. A car engine develops mechanical energy in the associated motion of its components. However, a stretched catapult also possesses mechanical energy – although stationary, it has the potential for movement, by virtue of the tension in the elastic. That is, when released the catapult will propel its payload causing it to move at great speed. Anything that is moving or has the potential to move can be said to possess mechanical energy (Figure 1.5), and we can further classify mechanical energy in terms of **▼Kinetic and potential energy▲**.

'Macroscopic' simply refers to a scale that is larger than 'microscopic'. In essence, it refers to events, objects and scales in the everyday world as opposed to those of microscopic and atomic phenomena.

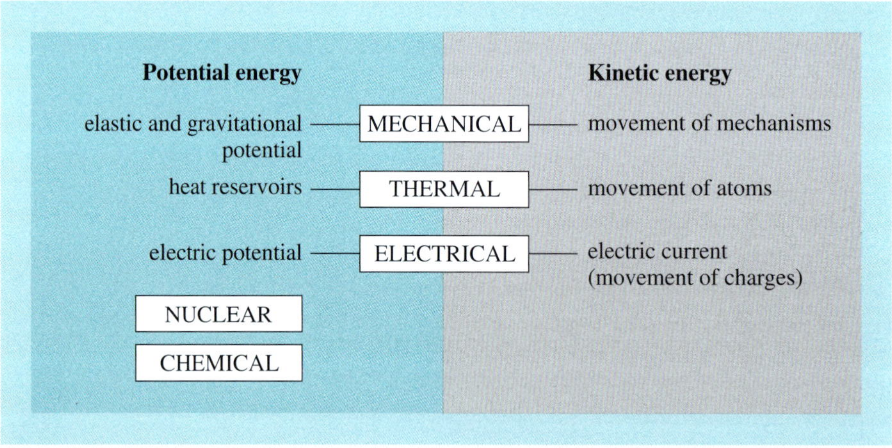

Figure 1.5 Forms of energy and their classification

▼Kinetic and potential energy▲

Anything that is either already moving (or would move if only you released the catch, let it fall, or opened the valve holding it back) has mechanical energy. We use the classifications *kinetic energy* and *potential energy* to distinguish between the energy associated with actual movement and that associated with the potential for movement.

Kinetic energy

Kinetic energy can be described as being either *translational* (where the movement is in a straight line) or *rotational* (where the movement is circular). For example, a car driving along a straight stretch of motorway possesses translational kinetic energy. However, its wheels and various engine parts are also spinning, so they possess rotational as well as translational kinetic energy. You should know that the translational kinetic energy of an object of mass m, which is moving with speed v, is given by:

$$K_{\text{trans}} = \tfrac{1}{2}mv^2$$

You should also know that the rotational kinetic energy of an object, that is rotating with angular speed ω and has a moment of inertia I about the axis of rotation, is given by:

$$K_{\text{rot}} = \tfrac{1}{2}I\omega^2$$

The total kinetic energy of an object, for example one of the wheels of a moving car, is the sum of its translational and rotational kinetic energies:

$$K_{\text{total}} = \tfrac{1}{2}mv^2 + \tfrac{1}{2}I\omega^2$$

Potential energy

The types of mechanical potential energy which we will consider here are *gravitational potential energy* and *elastic potential energy* (sometimes called *elastic strain energy*). On Earth, gravitational potential energy is the energy stored in an object as the result of its vertical position (i.e. height). Anything that has somewhere to fall under the influence of gravity has gravitational potential energy. Elastic strain energy is the energy stored in elastic materials as a result of stretching or compressing. Elastic potential energy can be stored in rubber bands, bungee cords, trampolines and stretched springs. The amount of elastic potential energy stored in each case is related to the amount of stretch (or tension) – the more stretch, the greater the stored energy. Equally, elastic potential energy can be stored in the compression of springs and of gases. Compression and tension are both involved in the potential energy of any component that is in a state of elastic bending (e.g. the archer's bow and the leaf springs on carriages were developed because they were useful reservoirs for potential energy).

The gravitational potential energy of an object of mass m, which is at a height h above the ground, is defined as follows:

$$U_{\text{grav}} = mgh$$

The elastic potential energy S of a spiral spring with spring constant k, which is extended a length e from its relaxed length, is defined as follows:

$$S = \tfrac{1}{2}ke^2$$

2.2.2 Electrical energy

When we plug an electrical appliance into a wall socket, we expect the socket to supply energy. This energy comes from somewhere else and is transported to the wall socket via a network of wires. But how does electrical energy travel along the wires? The answer is in the drift of electrons within the wires. In metals, electrons are free to move throughout the material and will do so if an electromotive force (e.m.f.) is applied. Electrons each have an electrical charge, and when they drift, this movement of charge is what is known as current. A mechanical analogy for electricity is water moving through pipes. Instead of an e.m.f., the water is driven by pressure, perhaps from a raised tank or a pump. The equivalent of electric current is the flow rate of the water. Just as moving water transports energy along the pipe, so does an electric current through the wire.

Electrical energy can be divided into equivalents of potential or kinetic energy, as in mechanical systems. If an electrical voltage is present (strictly, we speak of an electrical potential difference that is measured in volts) there is a possibility for electric current to flow; the only other thing needed is an electrical circuit for the current to flow in. In both electrical and mechanical systems, when we want to store energy it is usually in the form of potential energy, and when we are using it it is generally in some sense associated with motion, i.e. kinetic (or its electrical equivalent – a flowing current).

Exercise 1.3

(a) Mechanical energy can be stored for later use. Complete Table 1.3, which lists mechanisms for storing mechanical energy.

(b) When electrical energy needs to be stored, the stored energy is not necessarily electrical but may be in some other form that is easily converted to electricity within the storage module. Complete Table 1.4, which lists devices for storing energy that can be directly linked to electrical current.

Table 1.3 Mechanical energy stores

Storage mechanism	Release mechanism	Energy store
Watch spring	Balance wheel and escapement	
Water tower		Gravitational potential
Compressed air cylinder		Elastic potential
	Clutch and gearbox	Rotational kinetic

Table 1.4 Electrical energy stores

Storage device	Release mechanism	Energy store
Capacitor	Switch	
		Chemical potential
Flywheel-motor-generator	Switch or clutch and gearbox	

2.2.3 Chemical energy

All materials are made up of atoms and molecules. In all solids and liquids, and in most gases, atoms are bonded together in various ways, all of which are somehow to do with the interactions of electrical charges within and around the atoms. To break these interatomic bonds requires energy. Conversely, energy is released when interatomic bonds are formed to create molecules and larger assemblies of atoms. This energy is chemical energy. A large part of the study of chemistry is to do with ways of manipulating the arrangement of atoms through processes that involve much interchanging of energy, principally between chemical and thermal (but also often electrical) forms. Chemical potential energy is stored in the bonds between atoms.

2.2.4 Nuclear energy

Nuclear energy is similar to chemical energy in that it is associated with bonds between particles. In the case of nuclear energy, however, it is the bonds holding together the protons and neutrons within the nucleus which are important. To break and form these nuclear bonds requires much more energy than is needed to break interatomic bonds. Large amounts of energy are involved when the arrangements of protons and neutrons in atomic nuclei are changed. This energy is nuclear energy.

2.2.5 Thermal energy

When we put heat (thermal energy) into, or take it out of, an object, we see the effect as a change in its temperature. This can be expressed mathematically by saying that for every joule of thermal energy ΔQ put into an object of mass m, its change in temperature ΔT is given by:

$$\Delta T = \frac{\Delta Q}{mc}$$

where c is the specific heat capacity of the material the object is made from. The larger a material's specific heat capacity, the less its temperature changes when you put thermal energy into it.

Rearranging the equation, we can express a change in thermal energy content of an object in terms of its temperature change:

$$\Delta Q = mc\,\Delta T$$

We talk here about *changes* in thermal energy because, in practical engineering and real life in general, we are not interested in the total thermal energy, just the differences between objects at different temperatures.

In Block 1 Part 2 the thermal energy was linked to absolute temperature. On a microscopic scale thermal energy is manifested in the motion of atoms. In a solid, it is the vibrations of the atoms in the material that determines how much thermal energy it has and hence its temperature. The more vigorous the vibration, the greater the thermal energy and the hotter the material. The specific heat capacity then measures the extent to which thermal energy can be stashed away among the atoms.

Heat is the 'lowest' form of energy in that it is where all energy ultimately ends up. As heat, energy is highly disorganized, spread among atomic, microscopic and macroscopic systems.

2.2.6 Other forms of energy

Energy can be found in places other than the mechanisms and materials so far described. Don't be surprised to find energy associated with magnetism, with sound or with light. Whatever it is associated with, energy represents a capacity for work and for us an important question is the extent to which energy in a particular form is available for a particular task.

SAQ 1.2 (Outcome 1.3)

(a) Name five distinct forms of energy.

(b) Give one-sentence descriptions of each of your five named forms of energy.

Exercise 1.4

Identify the types of energy associated with the following things:

(a) a hot cup of coffee

(b) a lump of coal

(c) an incandescent lamp.

2.3 Energy conversion

Energy can be transformed from one form to another. For example, within our own bodies energy changes are vital. During metabolism, the chemical energy within food is converted to thermal energy to keep us warm, to electrical energy in the signals within our nerve fibres, and to mechanical energy in our muscles to enable us to move. Energy transformations take place all around us, all the time. However, although energy itself can be changed from one form to another, the total amount of energy always remains constant. Energy cannot be created or destroyed. This is the *Principle of Conservation of Energy*. As a result, energy is something you can account with (see **▼Balancing the energy books▲**).

▼Balancing the energy books▲

I expect that you have already got considerable experience of energy accounting. Tasks associated with keeping warm or keeping cool easily demonstrate that energy can be tracked joule by joule into or out of its thermal form. Back in Section 1 you have already been expected to use the idea that energy in one form at some time can be traced to an equal amount in some other form at some later time. The Principle of Conservation of Energy has some bearing on almost everything we do, although it is not always so apparent as in those examples.

The discussion of flight in Part 4 of Block 2 also used the idea of energy accounting in the Bernoulli equation. There, the mechanical energy of a flowing fluid was shown to be fairly transferable between a type that is kinetic (fluid flow) and others that are recognizably types of potential energy (pressure and gravity).

You have also, perhaps unwittingly, been involved with energy accounting in Block 2. When a beam is subjected to bending, energy is stored in the changed shape. In fact, the deflection under a given load can be calculated by detailed accounting for the work done in stretching and compressing the material as the beam deforms; see Figure 1.6.

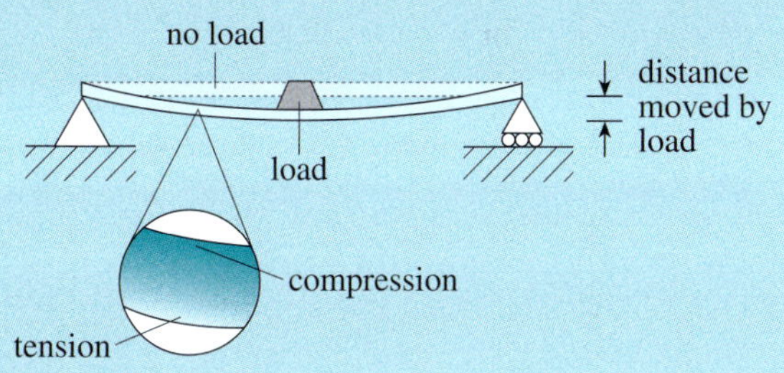

Figure 1.6 Potential energy lost from an object as it is rested on a beam that deforms under the load is transferred into compressed and stretched areas within the beam

Elastic bending means that the original mechanical energy can be recovered. So, for example, beams are used in relatively simple suspension systems – curved beams that deflect elastically as the load varies (Figure 1.7). Permanent deformation, however, puts the energy beyond reach. That too can be put to use, for example in the crumple zone of a motor-car body that is designed to absorb energy associated with impact.

Figure 1.7 A vehicle with curved-beam suspension

If you begin with a certain amount of energy and do something with it, then you have to be able to say where it all goes. In the case of a motor car, the initial energy is the chemical energy of the fuel. In an internal combustion engine, this energy is converted via thermal energy to mechanical energy. But not all of the chemical energy ends up as mechanical energy. The amount of mechanical energy produced by an internal combustion engine is less than the amount of chemical energy released from the fuel. So, where does the difference go? It can't just disappear.

Part 3 of this block will investigate the loss of energy in bearings.

The answer is that not all the thermal energy is retained within the car engine. The hot gases produced when the petrol is burnt lose heat to the air surrounding the engine. In addition, heat generated due to aerodynamic drag, and due to friction in the bearings and between the tyres and the road, is also lost to the atmosphere.

Whenever energy is transformed, the total output energy must be equal to the total input energy. However, during the energy conversions, some thermal energy often leaks away, so not all the final energy is necessarily available as useful energy. The ratio of the useful output energy to the total input energy is defined as the *efficiency* of the transformation process. It is a measure of the degree of success in transforming one sort of energy to another. Maximizing the efficiency of energy conversions is an important part of engineering and provides a major challenge for engineers.

Exercise 1.5

Identify the energy transformations that take place during the following events:

(a) winding up a clockwork toy

(b) a hot-air balloon taking off

(c) switching on a torch

(d) applying the brakes when cycling along a flat road.

2.4 Efficiency

Perpetual motion is an impossible thing – mechanisms always run down in the end, and we have to put in more energy to get them going again, whether by refuelling, or just winding up the spring again. Mechanical energy is somehow being lost from the system, ending up as thermal energy, whether we like it or not. Usually it is not difficult to find where it is lost, and we speak of friction, in the form of air resistance, lubricating oil viscosity, etc. It seems that every practical mechanical device we make will never achieve 100% efficiency. Inefficiency is similarly a part of more complex systems such as those that principally combine heat and motion (engines), or electricity and motion (motors and generators), or electricity and chemistry (batteries).

Efficiency can be quantified in terms of energy or power. The energy efficiency is linked to the achievement of useful work. For example, to pump some mass of water (m) up to a reservoir at a given height (h) consumes so many joules of electrical energy in doing so much useful work (mgh). The power efficiency represents an instantaneous measure. For instance for the same system, in pumping water uphill at a given rate, so many watts of electrical power are consumed. Whichever data are used to quantify the efficiency, the result should be the same.

In terms of energy the definition of efficiency is:

$$\eta = \frac{\text{useful energy output ('work done')}}{\text{total energy input}}$$

The Greek letter eta (η) is commonly used as the symbol for efficiency.

In terms of power the definition becomes:

$$\eta = \frac{\text{power output}}{\text{power input}}$$

Efficiency is quoted as a decimal number between 0 and 1 or, if multiplied by 100, as a percentage.

Many of our mechanical power sources work by converting chemical energy to heat and then to kinetic energy (think of an internal combustion engine). Any type of machine that works by converting thermal energy to mechanical energy (or vice versa) is technically known as a heat engine. Diesel engines, petrol engines, gas turbine (jet) engines, steam turbines and rocket engines are examples of heat engines. Machines that convert mechanical energy to thermal energy are also heat engines. A refrigerator is an example of a mechanical-to-thermal heat engine.

Heat engines have been studied theoretically in great detail, and a whole branch of science known as thermodynamics is devoted to the principles involved. You may recognize that the roots of the word combine the ideas of heat and motion.

The subject of thermodynamics was initially of interest in the design of steam engines. There is a great deal of choice in a steam engine as to how to generate the steam, what temperatures and pressures there should be in the various parts of the machine, and what kind of mechanism converts the potential energy stored in the pressure of the steam to kinetic energy. Nevertheless, by doing some inspired thinking, the thermodynamics experts of the nineteenth century reduced all this down to a surprisingly simple equation for the maximum possible conversion efficiency of thermal energy (input) to mechanical energy (output) of heat engines – see **▼On the efficiency of heat engines▲**.

The results of thermodynamics are of such fundamental importance that they are accorded the status of 'laws', and sit alongside those of Newton. The First and Second Laws of Thermodynamics are required to account for the energy changes and efficiencies of all heat engines (petrol, diesel, gas turbine, steam turbine and rocket). In fact the laws of thermodynamics are so fundamental that even electric motors have to conform, though you can still design a pretty good one without resorting to a thermodynamic analysis (and I doubt that such an analysis alone could lead to design improvements).

SAQ 1.3 (Outcome 1.6)

Identify which two of the following involve heat engines:

(a) electric motor

(b) deep freezer

(c) water wheel

(d) internal combustion engine.

▼On the efficiency of heat engines▲

The essence of any heat engine is that it takes energy as heat from a *source* (the 'hotter' reservoir in Figure 1.8), does some *work*, and then dumps some energy as heat to a *sink* (the 'cooler' reservoir). For example, in a steam engine, heat is generated by burning coal or some other fuel; some of this thermal energy is then converted to mechanical energy while the rest is transferred to the atmosphere.

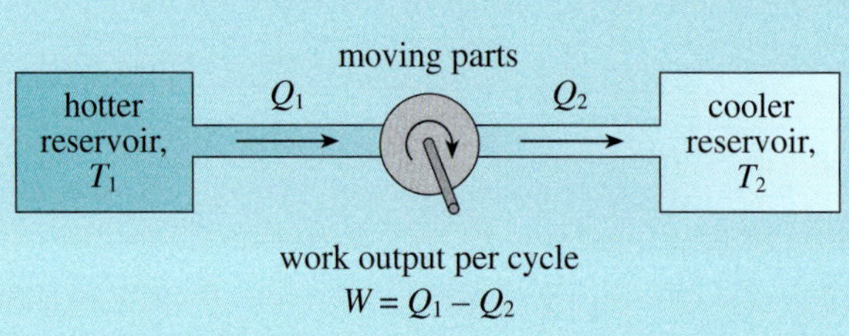

Figure 1.8 Schematic of a heat engine

Such an engine works continuously by operating in a cycle, with the engine repeatedly returning to its starting condition. The Principle of Conservation of Energy tells us that the best we can expect is that the work done per cycle is the difference between the energy that is put in, Q_1, and that which has to be dumped in the heat sink, Q_2:

$$W = Q_1 - Q_2$$

We can use this expression for the work done per cycle to help determine the efficiency of a heat engine:

$$\eta = \frac{\text{work done}}{\text{total energy input}}$$

$$= \frac{Q_1 - Q_2}{Q_1}$$

$$= 1 - \frac{Q_2}{Q_1}$$

As we established in Block 1 Part 2 and reiterated in Section 2.2.5 here, for a substance, thermal energy Q is proportional to its absolute temperature T. A heat engine generally involves a substance (such as water/steam or a fuel–air mixture) being heated, expanded and cooled. The efficiency can be written as:

$$\eta = 1 - \frac{T_2}{T_1}$$

where T_1 and T_2 are the input and output temperatures of the heat source and the heat sink.

The thermodynamic efficiency of a heat engine depends only on the input and output temperatures of the heat source and the heat sink:

$$\eta = 1 - \frac{T_2}{T_1}$$

In a steam engine, T_1 would be the temperature of the steam coming from the boiler, and T_2 would be its temperature at the exhaust of the engine. We can see that unless T_2 is at absolute zero, we can never attain 100% efficiency, but we can also see that we get most efficiency by making the difference between input and output temperatures as large as possible. The inventor Rudolph Diesel was fascinated by thermodynamics, and designed his engine to maximize this temperature difference.

Table 1.5 shows what typically happens to the energy that is put into a diesel engine. The data are for a six-litre engine fitted to an excavator. It is capable of delivering 67 kW of shaft power; this 'useful' output is only 34% of the input chemical energy of the fuel.

Table 1.5 Energy balance for a diesel engine on an excavator

Power income / kW		Power expenditure / kW		
from chemical energy in fuel	200	work to shaft against external load	67	(34%)
		heat to cooling water	46	(23%)
		heat lost in exhaust	58	(29%)
		all other heat losses	29	(14%)
Total	200	**Total**	200	(100%)

It is important to realize that the thermodynamic efficiency given here is the *absolute maximum* that Nature will allow. Frictional losses in bearings place additional limits on efficiency or, put another way, such losses contribute to inefficiency over and above the constraints of thermodynamics.

SAQ 1.4 (Outcome 1.6)

(a) In fewer than 30 words, state what is meant by efficiency.

(b) For a heat engine, give a mathematical expression that indicates the maximum efficiency that can be expected.

SAQ 1.5 (Outcome 1.6)

A power station burns 148 kg of coal every second. Each kilogram of coal contains 27 MJ of chemical energy. The electrical power output of the power station from its steam-turbine-driven generator is 800 MW. The inlet steam temperature to the turbine is 343 °C, and the temperature at the outlet of the turbine is 186 °C.

Calculate: (a) the total input power, (b) the operational efficiency of the station, and (c) its theoretical thermodynamic efficiency.

3 ELECTRICITY GENERATION

The generation of electricity involves transforming different types of energy into electrical energy. In the UK, the National Grid company buys from the power-generating companies and sells to regional distributors, coordinating a wide range of power sources through an aggressive trading market. The UK government publishes an official list of those installations that are licensed to supply.

A list of UK power stations and their generating capacities can be found on the T207 CD-ROM.

In fossil-fuel power stations, fuels such as coal, oil or gas are burnt (they undergo a chemical reaction with oxygen) to produce very hot gas that in turn heats a high-pressure boiler in which water is transformed into steam. The heated steam builds up a high pressure in the boiler, from where it is directed onto the blades of a turbine, forcing it to rotate. In effect the shaft is driven by a steam engine. It differs from the 'romantic' image of a steam engine chiefly in the use of a turbine to drive the shaft in place of pistons, connecting rods and cranks.

The turbine shaft is coupled directly to a generator that produces electricity (see **▼Power generation▲**). Here's the energy trail. The chemical energy stored in the fuel is converted to the thermal energy of the hot gas, from where it is passed to the skin of the boiler, then to the thermal *and* translational kinetic energy of the flowing, high-pressure steam, and from there to the rotational kinetic energy of the spinning turbine, and finally it passes via the generator into electrical energy – see Figure 1.9. Nuclear power stations work on a similar basis except that the heat required to boil the water is obtained through nuclear fission.

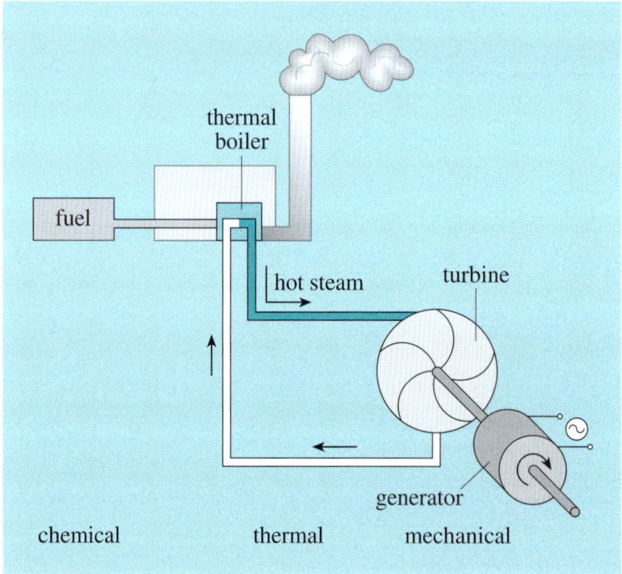

Figure 1.9 The energy trail from chemical fuel to electricity

Thermal power stations take too long to reach full power to be effective in dealing with sudden surges in demand. The reasons for this are: (a) it takes time for the fuel to heat the water to boiling point, and (b) it takes time for the steam to build up sufficient pressure for it to flow at a high enough speed to turn the turbine at the required rate. There is simply too much thermal inertia – too much matter has to have heat energy invested in it to get the system up to operating conditions – too much heat capacity and too much latent heat. In order to cope with sudden surges in demand another, more rapid means of getting the turbine rotating, and hence generating electricity, is required.

Latent heat is the energy investment that is required to bring about a phase change in a substance – here the transformation of water into steam.

Steam turbines are not the only way to drive a generator. Hydroelectric schemes use water to drive the turbines and 'gas turbines' are based on the aviation industry's jet engines, specially developed to deliver shaft torque rather than exhaust gas thrust. The most efficient gas turbine schemes for the generation of electrical power couple a steam cycle onto the output to extract heat from the exhaust – these are called 'combined-cycle gas turbines'.

The problem of how to supply electricity quickly at times when the demand suddenly increases is essentially one of how to bring about, efficiently, a rapid increase in the rotational kinetic energy of a turbine. For the rest of this Block, we will be taking a closer look at how this problem has been tackled.

▼Power generation▲

Figure 1.10(b) shows a schematic diagram of a turbine–generator assembly. The generator consists of a rotating 'rotor' and a stationary 'stator'.

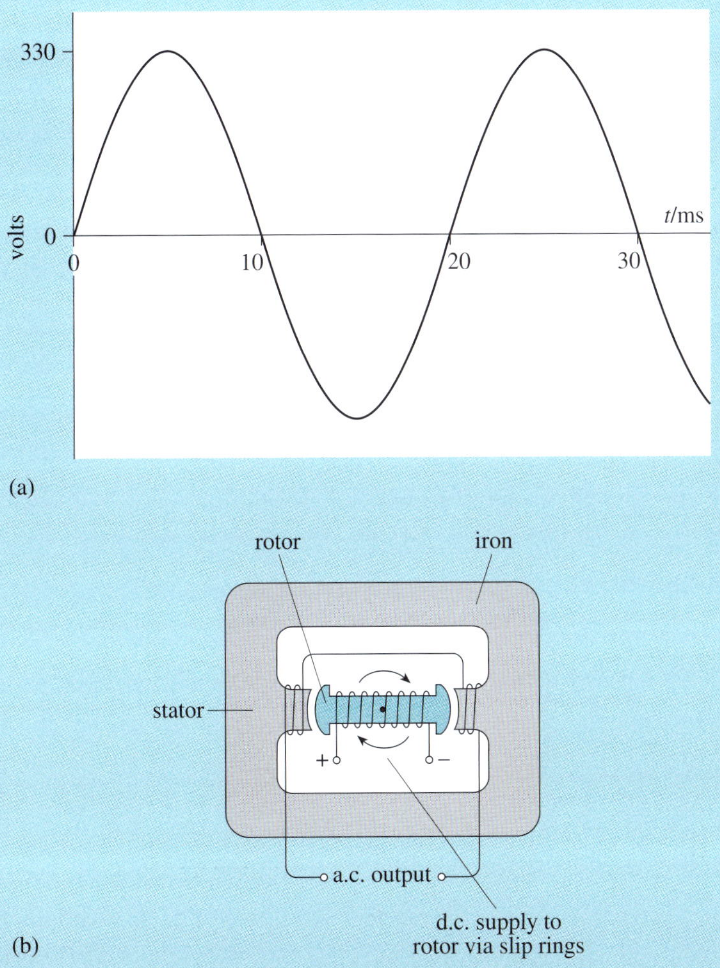

Figure 1.10 (a) Single-phase a.c.; (b) a simple 'single-phase' generator

The rotor is directly coupled to the turbine by a long shaft so that each time the turbine turns through one revolution, the rotor also turns through one revolution. Embedded in the outer ring of the rotor are a number of magnets – for serious power generation they are usually electromagnets. As the rotor turns, the poles of these magnets move past coils of copper wire wound on iron cores, together forming the stator. Moving magnets induce e.m.f.s in wires that will lead to current flowing in coils connected to external circuits (Block 3 Part 2). The direction of the current depends on the orientation of the magnetic field and its direction of motion relative to the wire. The magnets are arranged so that, as the rotor turns, first the north poles of the magnets pass the coils, then the south poles, then the north poles again, and so on – the magnetic field experienced by any particular part of a coil on the stator continually alternates as the rotor turns. The direction of the induced e.m.f. and any current that it causes to flow therefore alternate. The frequency of this alternating current (a.c.) is equal to the number of magnets (the number of pairs of N–S poles) embedded in the rotor multiplied by the number of revolutions that the rotor makes each second.

For usage in the UK, the standard a.c. frequency is 50 Hz. Accordingly, the product of the number of magnets and the number of revolutions of the rotor per second must equal 50. A rotor that is simply formed by a single magnet (one pair of N–S poles) must be rotated 50 times per second if the induced e.m.f. is to have a frequency of 50 Hz. A rotor that contains two magnets (two pairs of N–S poles) must be turned $50/2 = 25$ times per second. Similarly, a rotor formed from six magnets (six pairs of N–S poles) must be driven at $50/6 = 8.33$ times per second if it is to generate at 50 Hz.

Industrial motors require so-called three-phase electrical power, and national distribution grids normally distribute power ready for this; domestic users are usually supplied with power from just one of the three phases as part of the final stage of distribution. You will have seen pylons carrying two sets of three lines, one on each side; the three lines in a set form a three-power-phase circuit. So the pylon holds six separate lines, suspended from six separate insulators. In a three-phase system there are three wires, each carrying 50 Hz alternating current but fed in such a way that the current in any one wire lags exactly one-third of a cycle behind that in one of the others and leads that in the third wire by a similar amount – see Figure 1.11.

If you look carefully at the tallest pylons you'll probably notice that each of the six insulators actually supports a bundle of four wires. That's a technicality associated with high-voltage transmission. All four wires in a bundle make up the same phase.

To generate the three phases the generator is equipped with three sets of coils displaced around the stator by 120° between coils. This ensures that the spinning of the rotor excites the coils appropriately.

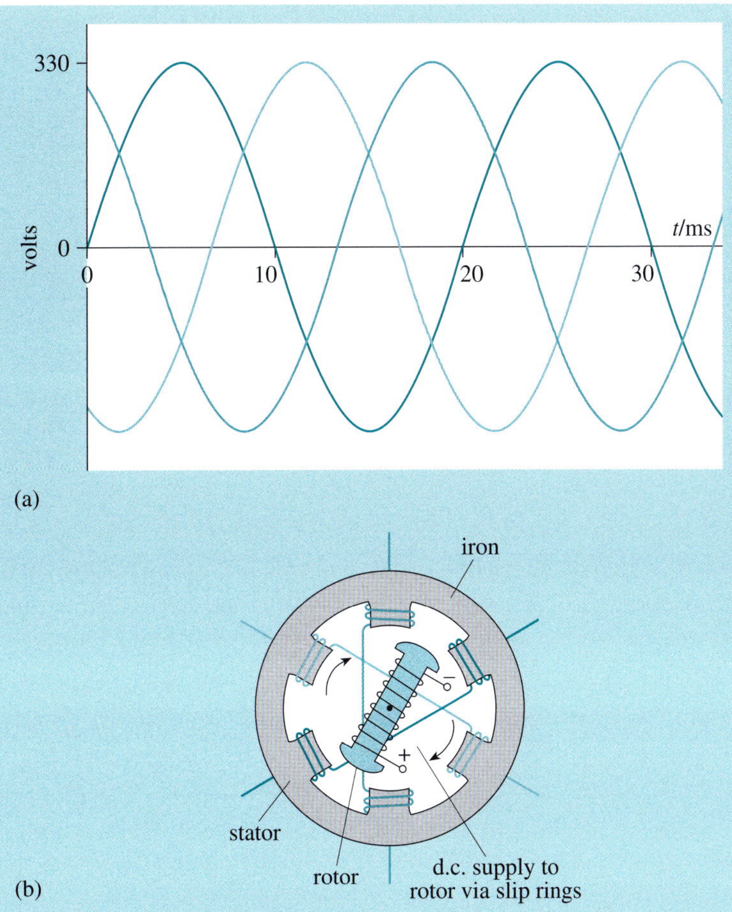

Figure 1.11 (a) Three-phase a.c.; (b) a simple 'three-phase' generator

4 ELECTRICAL ENERGY STORAGE SCHEMES

Electricity generation and distribution on a national scale is very much a game of minimizing both capital and running costs and maximizing efficiency. We have already seen that the demand for electricity can vary enormously, and sometimes very rapidly too. So how has the system been organized to cope with this greatly fluctuating demand at the lowest possible cost?

4.1 Exploring the problem

Let's follow the strategy outlined in Part 1 of Block 1 to investigate storage schemes for the electricity supply industry. The first thing to do is to establish the 'need', to ensure we have enough background information to specify the problem that is to be solved.

Electricity generation plant is arranged on a hierarchical basis: to supply the 'base load' – the level of demand below which we never go – we have several very large, fossil-fuelled power stations and the nuclear power stations. Here the large capital cost of building them and the long start-up and shut-down times mean that it's best to run them continuously. As demand increases, less efficient and smaller power stations are brought on stream. In practice, considerable planning underlies the strategy, taking account of weather patterns and forecasts, in conjunction with virtually continuous negotiations between station operators and the grid controllers. The whole thing is driven by a broader commercial energy market that trades throughout the day and night.

In order to further shorten the start-up time for generating extra power, the National Grid maintains some power stations in a standby mode, with the turbines spinning at exactly the right rate and synchronized with the 50 Hz a.c. supply, but with little electrical energy actually being drawn from them. In the industry, they call this 'spinning reserve'. In addition a standing reserve of generating capacity is maintained at 'less than twenty minutes' notice'.

There is room in this system for a few large reserves of energy that can be called upon quickly when supplies are short, and that can then be recharged *electrically* when supplies are plentiful. This is the particular 'need' for which solutions are required.

4.2 Specifying the challenge

A functional specification can be devised for a scheme to meet the need for standby energy. It is likely to be based around the standard generator technology, so let's restrict the specification to one that is. Exercise 1.6 will start you thinking about what has to be specified, and SAQ 1.6 will test your ability to quantify and extend the specification.

Exercise 1.6

Look back at the data for the TV pickup shown in Figure 1.2 and suggest *two* quantities that should be included in the performance specification for an electrical energy reserve that might be called on to maintain supplies throughout such an event.

SAQ 1.6 (Outcome 1.4)

(a) Complete Table 1.6 with numerical data based on the case illustrated in Figure 1.2.

(b) Suggest two factors that might be considered in deciding where an electrical energy reserve could be located.

Table 1.6 Functional specification for a standby energy scheme

Generator technology	Standard three-phase
Power / MW	
Response time / s	
Energy / GJ	
Recovery time / h	
Recharge technology	Electrically driven

4.3 Opportunities

4.3.1 Possible solutions

According to Block 1 Part 1, what should happen now is that we look for possible solutions in outline and then we start to evaluate them; so here goes.

Example

Some kind of hydroelectric scheme has been anticipated throughout the text. Outline the scheme in the light of the specification summarized in the answer to SAQ 1.6.

Solution

A 2000 MW hydroelectric scheme working between two reservoirs pumps water up when supplies are plentiful and generates electrical power in water-driven turbines when electrical supplies are short.

Now it's your turn as a member of the problem-solving team to put forward an idea for a solution via the following exercise – if you want some inspiration check through the existing supply base on the T207 CD-ROM.

A list of UK power stations and their generating capacities can be found on the T207 CD-ROM.

Exercise 1.7

Without restricting your thoughts by judgements on practicality, outline at least one other potential solution to the specified need that is summarized in the answer to SAQ 1.6.

4.3.2 Evaluating solutions

The next step would be to sort out the feasible ideas from the over-imaginative. In this case that means we should devise models that can be used to check the feasibility in quantitative terms. I want to concentrate on one solution to illustrate this stage of the process.

The answer to Exercise 1.7 proposed other possible solutions, some of which are discussed further in Section 6.

One of the solutions that emerged above, that outlined in the example, is one that has been successfully implemented in the UK and it involves storing energy in the form of the gravitational potential energy of water in a high reservoir. At times of high demand, this gravitational potential energy is rapidly converted to the rotational kinetic energy of turbine–generator assemblies to generate electricity. Equally importantly, at times of overcapacity, this system can buy surplus electrical power cheaply to pump water up to the storage reservoir. These are the principles of *pumped storage*.

In the next section, we'll go through some of the evaluation steps based on the design that is currently in operation at Dinorwig, North Wales – one of the largest pumped storage schemes in the world. We know that the Dinorwig installation works satisfactorily, so it will be interesting to see if our design evaluation is good enough to have kept the idea alive had we been solving this problem 'for real'.

5 THE DINORWIG PUMPED STORAGE SCHEME

Between 1974 and 1982 at Dinorwig, in North Wales, a huge pumped storage scheme was built. It contains Europe's largest man-made cavern and 16 km of tunnels, and its construction required one million tonnes of concrete, 200 000 tonnes of cement and 4500 tonnes of steel. However, there is little to see as you drive past, as most of it is deep inside Elidir mountain. The only visible evidence of the pumped storage power station is two reservoirs, one located part of the way up the mountain and one at the bottom.

Water is pumped up to the upper reservoir (Marchlyn Mawr) using electrical power. This is normally done at night when the demand for electricity across the country is low. Whenever there is a sudden demand for power, the headgates (huge taps) at the top reservoir are opened. Water rushes down a set of tunnels to six turbines, which in turn drive powerful generators. The water then collects in the lower reservoir (Llyn Peris), ready to be pumped back up later. Figure 1.12 shows a graphic visualization of Dinorwig power station. The key components can also be seen in the schematic cross-section shown in Figure 1.13.

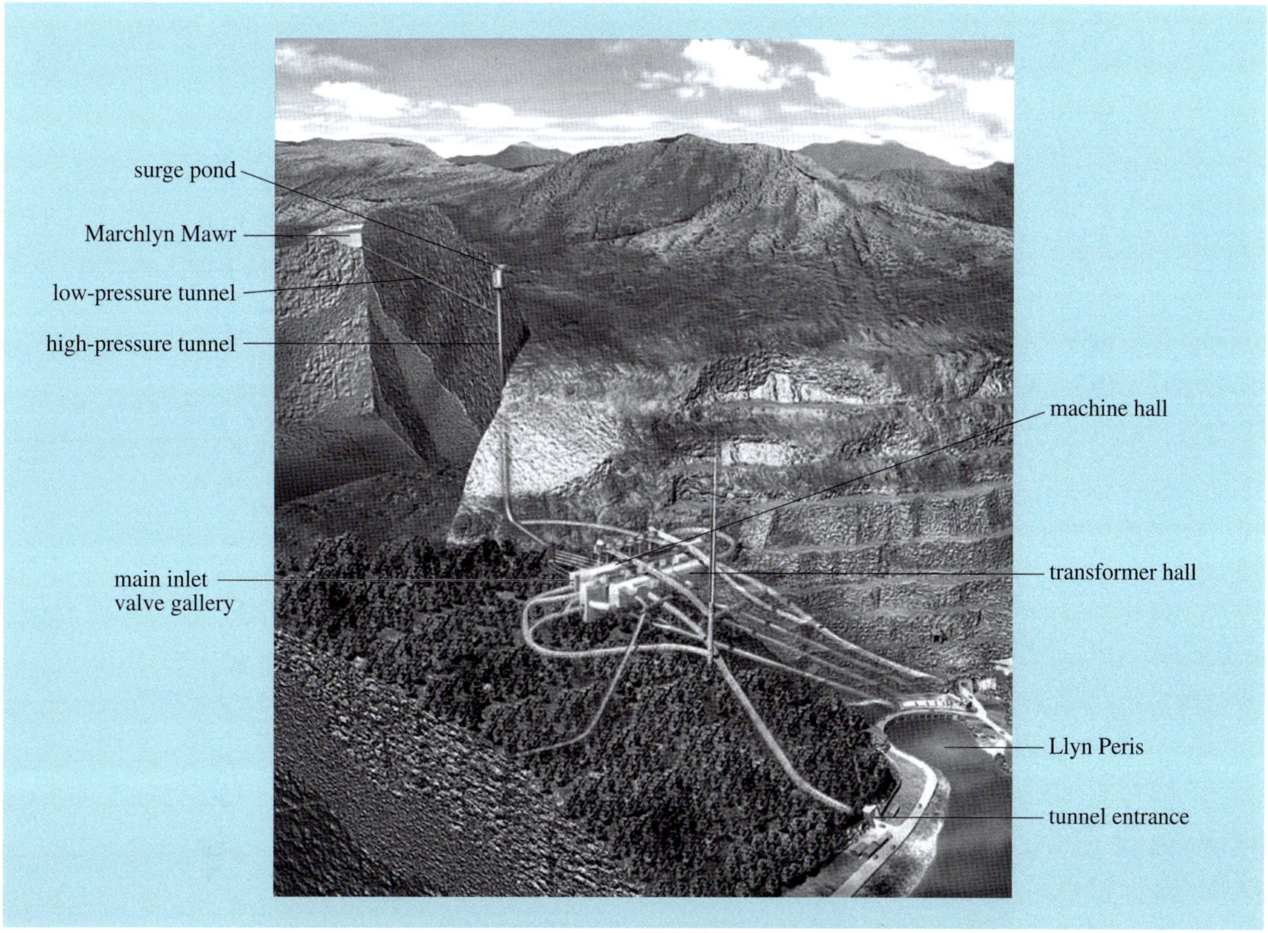

Figure 1.12 The Dinorwig scheme: all the machinery and pipework is hidden away within the mountain so that the visual impact is negligible compared with that of most power stations of equivalent capacity

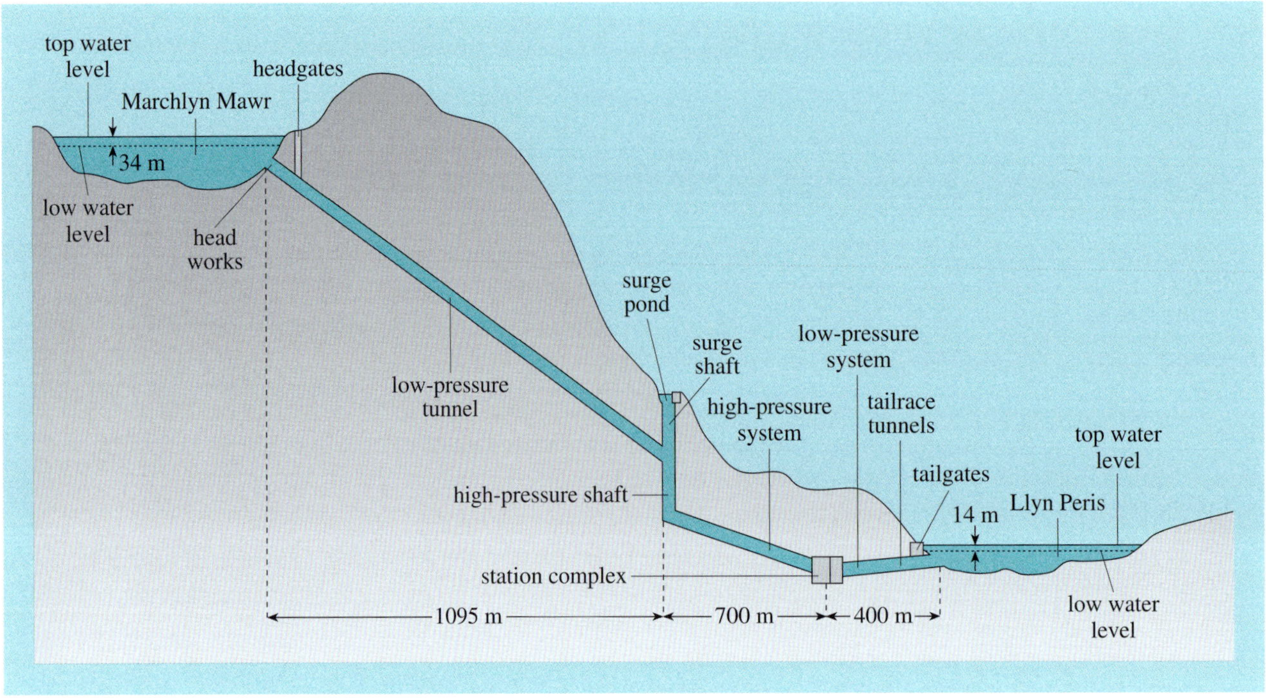

Figure 1.13 Schematic cross-section of Dinorwig scheme. The principal components, other than the reservoirs and tunnel entrances, are hidden within the mountain

When water is pumped up to the upper reservoir, the electrical energy used in the pumping is converted into the gravitational potential energy of the water. The greater the height, the more energy is stored. This is one of the reasons that the Dinorwig site was chosen – the height difference between the two natural lakes, that were enlarged to form the Marchlyn Mawr and Llyn Peris reservoirs, is over 500 m. The opening of the headgates results in the conversion of the gravitational potential energy to pressure potential energy and translational kinetic energy, as the water flows through the low- and high-pressure tunnels down the hillside. At the end of the high-pressure tunnel, the water is directed along steel-lined pipes to six turbines. The fast-moving water causes the turbines to turn, converting the input energy to rotational kinetic energy. Finally, the generators convert the rotational kinetic energy to electrical energy.

There are six generating units at the Dinorwig power station and each is a reversible pump–turbine. That is, as well as acting as a turbine-driven generator, the hardware functions also as a motor-driven pump, to move water from the lower to the upper reservoir.

5.1 Power and energy

Stand by with your calculator and a notebook, as I am going to ask you to work through some arithmetic. To estimate the power output of the Dinorwig power station we need some basic data concerning the site and the installation. Table 1.7 gives information about the heights of the two reservoirs, the volume of water that can be transferred between them, and the rate at which that water is transferred.

Table 1.7 Reservoirs in the Dinorwig scheme

Volume of water available for transfer between the two reservoirs	7 000 000 m^3
Rate at which water is transferred from upper to lower reservoir	390 $m^3\ s^{-1}$
Upper reservoir maximum water level (height above sea level)	634 m
Upper reservoir water level range	34 m
Lower reservoir maximum water level (height above sea level)	106 m
Lower reservoir water level range	14 m

SAQ 1.7 (Outcomes 1.5 and 1.7)

Use the data in Table 1.7 to do the following:

(a) Calculate the mass of water that is available for transfer between the two reservoirs. The density of water is $\rho = 1000$ kg m^{-3}.

(b) Calculate the average water level (as a height above sea level) of the upper reservoir.

(c) Calculate the average water level (as a height above sea level) of the lower reservoir.

(d) Find the average height through which the water drops.

(e) Calculate the available gravitational potential energy if the upper reservoir is full.

(f) Calculate how long it will take for the upper reservoir to empty into the lower reservoir.

(g) Calculate the expected power output during the five hours for which the station can operate.

From your calculations, you should have found that around 35 600 GJ of gravitational potential energy can be stored when the upper reservoir at Dinorwig is full, and that it can release that energy over a period of five hours. The expected power output is then approximately 1980 MW. So far, so good.

5.2 Efficiency

The six generators at Dinorwig are each capable of producing 288 MW of power at 18 000 volts, which is stepped up to 400 000 volts by transformers and fed into the National Grid.

SAQ 1.8 (Outcomes 1.5, 1.6 and 1.7)

(a) Each of the six generators at Dinorwig can produce 288 MW of power when all of them are driven simultaneously by the maximum possible flow of inlet water. Calculate the total output capacity of the Dinorwig pumped storage power station.

(b) Compare your answer to part (a) with the predicted power output of the station calculated in SAQ 1.7 (g).

(c) What is the efficiency of the Dinorwig power station expressed (i) as a decimal, and (ii) as a percentage?

Hydroelectric power stations are generally between 80% and 90% efficient. This is high compared with fossil-fuel and nuclear power stations, which commonly have efficiencies in the range 40 to 60%. The remainder of the original energy in the fossil or nuclear fuel ends up as waste heat.

In hydroelectric power stations, we are not trying to convert thermal energy to electricity, and we are therefore not constrained by the thermodynamic considerations set down in the thermal efficiency equation. Where efficiency is crucial, it is wise to avoid thermal cycles if possible.

This efficiency of hydroelectric power generation is primarily that of the turbo-generating apparatus. The rain that feeds the reservoir is a result of solar energy evaporating water from the ocean and lifting the vapour high into the atmosphere, before it re-condenses and falls as rain. After passing through the turbines the water is discharged into a river. In the case of Dinorwig, water discharged from the turbines is kept in a lower reservoir from where it is pumped to the upper reservoir. The pumps use electricity generated by other (predominantly thermal) power stations.

There are three efficiencies that need to be considered. For the economists it is the efficiency of the pumps and that of the turbo-generators that is important. Environmentalists might be more concerned with the efficiency of the primary source. Dinorwig does not really make 'new' electricity; it simply delays the use of that already generated. In so doing it can enable more efficient or more economical use of the primary generating capacity.

SAQ 1.9 (Outcomes 1.5, 1.6 and 1.7)

(a) The Dinorwig hydroelectric power output, according to my answer to SAQ 1.8, is 87% of the power input. The energy input was found in SAQ 1.7 to be 35 600 GJ. Use these figures to estimate how many gigajoules of energy are lost (in the sense that the electrical output is less than the gravitational potential energy input) when all the available water is transferred from the top reservoir to the bottom reservoir.

(b) Think carefully about the different stages through which the gravitational potential energy of the water is transformed into electrical energy. Suggest where and why energy is lost.

Bearings are the subject of Part 3 of this block.

Most of the energy lost in converting the gravitational potential energy to electrical energy at Dinorwig ends up as heat. In particular, frictional losses arise in the bearings that support the weight of the turbo-generator while allowing its shaft to be driven. Frictional losses also occur in the interaction between the flowing water and the tunnels, pipes and channels through which it is guided.

5.3 Response time

Despite the energy losses, Dinorwig still has a capacity of 1728 MW. This is large enough to make it useful in helping cope with the magnitudes of surges in UK demand caused by TV pickup. However, is Dinorwig able to respond quickly enough to such surges? In this section, we'll try and calculate just how long it takes for turbines like those at Dinorwig to be cranked up to full power.

What we need to find is how long it takes for the flowing water to get the turbine–generator assemblies turning fast enough to produce electricity for use by the National Grid. After that there will be a short interval during which the speed of each generator is finely adjusted before it is synchronized with the grid ready for connection.

A rough estimate

Let's take a closer look at one of Dinorwig's six turbine–generator assemblies. We'll start off by neglecting any friction losses in the bearings and we'll assume that the generator is not connected to the grid. In other words, we'll try and calculate how long it takes to get the mass of the machinery rotating at a given speed.

Table 1.8 gives basic information about the turbine–generator assemblies including their moment of inertia, calculated from the dimensions and mass of the actual units. If these data are not available some further modelling would be required to link a particular generating capacity with a particular size and shape of machinery.

Table 1.8 Data on Dinorwig turbine–generator assemblies

Number of turbine–generator assemblies	6
Moment of inertia of each turbine–generator assembly	1200×10^3 kg m^2
Turbine inlet pipe area	4.2 m^2
Effective radius of turbine at inlet	1.9 m
Height of turbine inlet pipe (above sea level)	32 m

To be accepted by the National Grid, the electricity generated must have a frequency between 49.5 Hz and 50.5 Hz. The turbine–generator assemblies at Dinorwig each contain six magnets. Therefore, to provide an a.c. current of 49.5 Hz, the rotor must turn $49.5 / 6 = 8.25$ times a second, or through 51.84 radians per second. So, we are looking to find out how quickly the turbine–generator assembly can be made to rotate at 51.84 rad s^{-1}.

As we are trying to establish how long it takes for the turbine–generator assembly to reach a given angular speed, what we need to find is its angular acceleration. You should recall from Block 3 Part 2 that when a torque is applied causing something to rotate, the torque is equal to the moment of inertia multiplied by the angular acceleration:

$$\Gamma = I\alpha$$

Rearranging gives the following expression for the angular acceleration of the turbine driven by a particular torque:

$$\alpha = \frac{\Gamma}{I}$$

The moment of inertia I for one of Dinorwig's turbine–generator assemblies is given in Table 1.8 as 1200×10^3 kg m^2, so to work out the angular acceleration of the turbine we just need to find the applied torque. Rather than trying to work out the available torque from first principles we can take a short cut and use the parameters that have already been established for what the system must be doing if it works. So, we know that a turbine will generate about 288 MW when being driven flat out and we know that when it is idling, producing no electrical power, and when it is connected to the grid, it has to be running at an angular speed of 51.84 rad s^{-1}. The turbine manufacturers will also be able to confirm that frictional forces when idling off load amount to 2 MW. From these we can get estimates of the **▼Maximum torque and standby torque▲**.

'Idling' is used here to imply a system that is running in a standby mode that delivers no useful power.

▼Maximum torque and standby torque▲

In Block 3 Part 1 it was established that a torque, Γ, that turns a shaft at a speed ω transfers a mechanical power of:

$$P = \Gamma\omega$$

That is, the rate of change of angular speed arises from the difference between the driving force and a friction or drag force, which in this case is presumed to be the common type of drag friction that is proportional to speed.

Standby torque

When idling on standby (2 MW) at the minimum design speed of 51.84 rad s^{-1}:

$$\Gamma_{\text{standby}} = \frac{2 \times 10^6\ \text{W}}{51.84\ \text{rad s}^{-1}} = 0.04\ \text{MN m}$$

Maximum torque

At full power the shaft overcomes friction and delivers power to the generator. Neglecting generator inefficiency when running at the minimum design speed:

$$\Gamma_{\text{full load}} = \frac{(2 + 288) \times 10^6\ \text{W}}{51.84\ \text{rad s}^{-1}} = 5.6\ \text{MN m}$$

The angular acceleration if the gate valves are opened fully is therefore:

$$\alpha = \frac{\Gamma}{I} = \frac{5.6 \times 10^6\ \text{N m}}{1200 \times 10^3\ \text{kg m}^2} = 4.7\ \text{rad s}^{-2}$$

That is a massive acceleration and it will spin the turbine from rest up to the design speed in about 12 seconds, provided the machinery can stand that kind of rough treatment (which I doubt); see Figure 1.14.

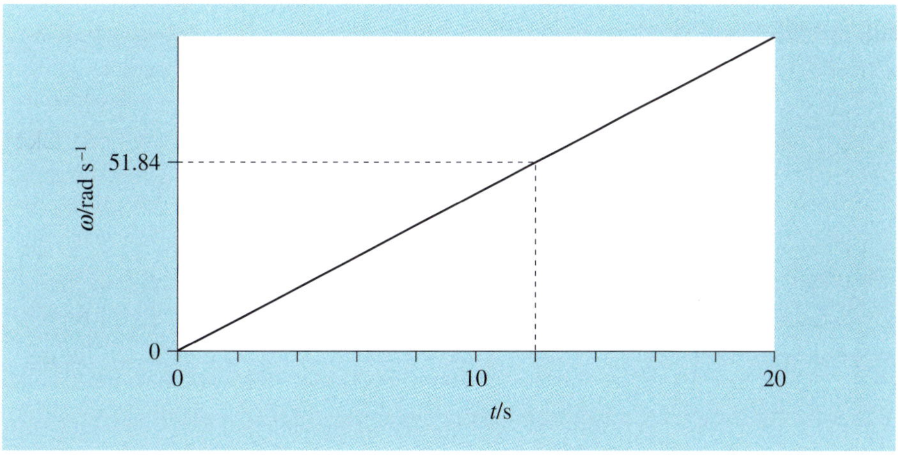

Figure 1.14 Acceleration of the turbine at full power

A gentler approach would be to set the inlet valves at the idle position, whereupon the acceleration from rest will be much gentler:

$$\alpha = \frac{0.04\times10^6 \text{ N m}}{1200\times10^3 \text{ kg m}^2} = 0.03 \text{ rad s}^{-2}$$

But at this rate it will be half an hour before it's up to speed; see Figure 1.15. This is not much faster than the response time of a fossil-fuel or nuclear power station. Ideally a faster response is required to cope with the surges in demand.

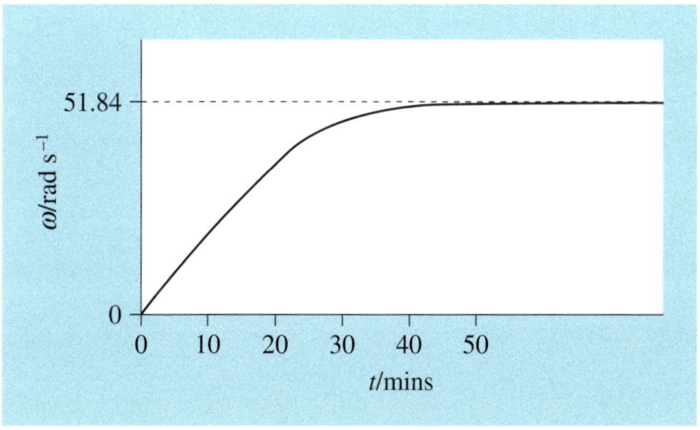

Figure 1.15 Acceleration of the turbine at standby power

What would you do to hurry things along? I would open the valve steadily until it was approaching half speed and then start closing the valve down again to the idle position. That's easily said and fairly easily done, but it's a bit harder to solve mathematically as the acceleration is clearly not constant.

The non-constant acceleration of a turbine is analysed in a *Mathcad* worksheet on the T207 CD-ROM.

What happens in practice

In fact, Dinorwig has the fastest response time of any pumped storage plant in the world and does actually provide 1320 MW of its available 1728 MW in just 12 seconds. The reason why it can do this without wrecking the hardware is that it doesn't start from rest. Normally the turbine is entirely immersed in water, but when a unit is on standby the water is replaced with an enormous air bubble provided by a large tank of compressed air – there is one for each generator. The machinery is then kept spinning on standby not by water flow but by running the generator as a motor, drawing electric power from the grid. When the instructions come to start generating, the main inlet valve opens within five seconds and the turbine, already spinning and synchronized in air, is then driven by the inlet water, generating power almost at once.

5.4 Recovery time

The necessary delay while the system is 'recharged' also appeared in the specification. For the Dinorwig scheme this is done by running the system backwards. The nature of electrical machines is such that there can be a smooth transition between generating and motoring when they are connected to a large supply network – if a generating machine is not driven hard enough it will consume power rather than produce it. By contrast, not all hydraulic machines are reversible: not all turbines will act as pumps when driven backwards. The turbine design for Dinorwig was selected to ensure that this could be done.

Turbines are the subject of Part 2 of this block.

Because the system can be operated in reverse, it is not surprising that the time to recharge the upper reservoir is only a little longer than that for which it is discharged. In this mode the system can make use of almost 2 GW of surplus capacity.

5.5 Dinorwig's contribution to dealing with UK demand

Earlier we took a close look at the UK's demand for electricity during events such as the England *v* West Germany World Cup semi-final of 1990 and the solar eclipse of 1999. Let's look at how the Dinorwig pumped storage station was used to help cope with this type of surge in demand.

SAQ 1.10 (Outcome 1.1)

Table 1.9 details the electricity generated by the Dinorwig pumped storage station during the morning of the solar eclipse on 11 August 1999.

(a) Plot a graph of output versus time.

(b) Compare the graph you plotted in part (a) with the graph of demand you plotted in answer to SAQ 1.1. The demand showed a surge of approximately 3000 MW. Estimate what percentage of this sudden electricity requirement was provided by Dinorwig.

Table 1.9 Dinorwig's output during the morning of the solar eclipse on 11 August 1999

Time	10:40	10:50	11:00	11:10	11:15	11:20	11:30	11:40	11:50
Output / MW	750	730	430	0	0	300	1130	1300	800

6 OBSERVATIONS ON ALTERNATIVE SCHEMES

Before the Dinorwig scheme could be built there were hundreds of secondary tasks for the problem solvers to tackle, and we'll look at just two areas in more detail in Parts 2 and 3 of this block.

The Dinorwig power station is certainly a success story, but it represents only one of a number of possible solutions to meeting the need for almost instant generation of large amounts of electricity. Let's take a look at some alternatives to see where their secondary challenges lie.

6.1 Compressed air

Mechanical energy can be stored by pressurizing a volume of air. The energy that can be stored in a compressed gas depends on how quickly it is compressed and how soon the energy is to be recovered. To evaluate this solution we could do a quick calculation to see how large a reservoir of air we would need in order to store an equivalent amount of energy to Dinorwig.

What we'd find is that to be made of steel or concrete above ground it would have to be extremely thick-walled (it would also be an incredibly dangerous potential bomb). We could perhaps find or excavate a large cavern and if necessary line it with some kind of airtight material. Even if we did, what makes this system particularly awkward to use is that the pressure of the air as it is extracted is continually falling, until the last bit is only just above atmospheric pressure. It would be difficult to design a turbine that would run efficiently over this entire pressure range (remember also that we need to maintain the electrical output at 50 Hz, so we need a constant rotation rate). This means that, in practice, not all the energy stored would take part in the charging and discharging of the store. Air would be drawn from the store until its pressure had fallen to, say, 50% of its maximum. This means that the effective capacity of the store is only a half of its actual energy content, and we'll never use half the energy we put in when it was charged for the first time.

Another factor to consider is that as gases are compressed, they heat up. If this heat is allowed to dissipate, it represents a loss of energy. What might be needed is some separate means of storing this thermal energy and returning it to the air when it is allowed to expand again on its way through the turbine, thus counteracting the cooling effect that also happens as a consequence of expansion.

With all these drawbacks, it may come as a surprise to learn that compressed-air power-storage schemes *have* been built. The first was commissioned as long ago as 1978 in Huntorf in Germany, and others have been brought on stream in Japan and the USA. The Huntorf reservoir is a natural salt cavern – the surrounding rock forms an excellent pressure vessel.

The Huntorf plant has an output of 290 MW, which it can sustain for two hours. Its chamber has a volume of 300 000 m^3, at a pressure of 75 times atmospheric. When you do the calculations, you find that 290 MW for 2 hours is 2000 GJ, but the *total* energy content of the air reservoir is only 2250 GJ. Yet we just said that we'd not be able to use more than about a half of the reservoir's capacity, so where is the extra energy coming from?

The answer is that, unlike pumped water storage schemes, compressed-air schemes are not linked to simple turbines and they also use fossil fuels. In a conventional natural gas turbine plant, nearly two-thirds of the output power is used to compress the air before it passes into the combustion chamber. With the air already compressed in a storage reservoir, the turbine does not have to drive a compressor. Instead, all its output torque is delivered to the generator, making it some three times more efficient. There's a lesson here about keeping an open mind when innovating.

6.2 Batteries

We use batteries everywhere to store energy in chemical form for later use in electrical form. Chemistry is all about moving electrons from one atom to another. If you can physically separate the atom giving up the electron from the one receiving it, and arrange things in such a way that the only way the electron can get from the one to the other is by going through an external circuit, then you have a battery.

Whenever chemical rearrangements occur, atoms are required to swap or share electrons with each other. Energy is bound up in any arrangement of atoms. Where the result of a rearrangement binds up less chemical energy than before, chemical potential energy is released. In some cases the energy is released as electrical energy, just as a rolling ball can transfer its potential energy into kinetic energy as it rolls down a hill. The cell within a battery is configured to intercept these transactions so that the electrical energy made available by the chemical reorganization can be gathered up. Ultimately, things are arranged so that there is a separating of electrical charge, driven by chemistry. You already know that we call the subsequent pushing and pulling of charges an electromotive force.

There are three components in any battery: two different conductors known as electrodes, and the conducting medium between them, which is the electrolyte. All batteries combine an electrolyte with pairs of dissimilar electrodes. If you want to generate something useful in the way of electricity then the electrodes must be of different materials.

A *primary battery* is designed to extract electrical energy from the reaction of a specific combination of chemicals, after which it is discarded.

A *secondary battery* is the technical term for what the general user calls rechargeable. The rechargeable battery is not a new idea. From the early days it was appreciated that some combinations of electrodes and electrolytes could be reinvigorated by pumping electric current back into them. This reverses the chemical reactions that take place in normal use, effectively recycling the component materials *in situ* (i.e. without taking the unit apart).

Secondary systems effectively store energy by converting it from electrical into chemical energy. In earlier times they were also called accumulators because of the way energy could be accumulated. In a secondary system energy can be stockpiled for subsequent use.

The rechargeable lead–acid batteries in every road vehicle continue to function for many years, and over many charging and discharging cycles. They are able to deliver considerable power, for instance to turn the starter motor; a typical lead–acid battery has a specific power of 100 W kg^{-1} with a specific energy of about 8 kJ kg^{-1}.

Could a storage scheme be designed around secondary batteries? Though not quite on the scale of the National Grid, diesel–electric systems have proved a popular alternative to nuclear energy for powering submarines, so the idea is certainly worthy of further evaluation, though we'll leave it for another day.

6.3 Kinetic energy storage

In Part 2 of Block 3 I described how to store energy as the rotational kinetic energy of a flywheel. The case considered was that of an 'uninterruptible' power supply used for ensuring the continued operation of computer systems during a mains failure. A flywheel is 'charged up' by an electric motor. When the mains supply is cut off the motor acts as a generator, recovering electrical energy from the spinning mass of the flywheel. The flywheel slows as it gives up its rotational kinetic energy. That kind of kinetic energy store was calculated to have a specific power of about 130 W kg^{-1} and a specific energy of about 324 kJ kg^{-1}.

The functional specification we constructed in Section 4 was looking for 2000 MW and 5400 GJ. Any volunteers for the design team?

7 PROBLEM SOLVED?

In this opening part of Block 4 I have outlined the need for a rapid response to electrical power demand on a national scale. In exploring the specification it was apparent that there might be several ways to fulfil this need. We evaluated the overall design of one solution based on the successful pumped storage scheme at Dinorwig. Interestingly, the need for rapid response has evolved since the original concept of pumped storage for coping with occasional extreme demand surges. When the UK electricity market was deregulated, hour-by-hour price fluctuations introduced opportunities for suppliers who could operate a policy of selling energy when the electricity price is high and buying it when it is low. Pumped storage is particularly adept in this market owing not only to its rapid response and reversibility, but also to its large capacity. Of course, the owner had to re-evaluate the system to ensure it could be operated in this new, fast-changing market.

We have not studied the whole of the Dinorwig project, taking for granted the blend of innovations and routine engineering that were combined by the original design and construction teams. Many hurdles stand between the outline feasibility of a large project like this and its implementation. Other challenges arise when the original need evolves, as has been the case here.

The Dinorwig project has had to cope with environmental issues as well as with major challenges in civil, electrical and mechanical engineering. For example, the specification of the pipework that has to carry 60 $m^3 s^{-1}$ of water at twenty times atmospheric pressure to the inlet of a turbine is not a trivial matter. We'll return to this particular challenge in the *Dinorwig Case Study*, after we've studied some of the ways to design against catastrophic failure in Block 5.

Turbines are at the heart of most electrical power generators. Block 4 continues by investigating water turbines. After that we turn our attention to one of the other areas highlighted in the story of large machines for electrical power generation: the bearings that support hundreds of tonnes of spinning hardware.

So, it's not really a case of 'problem solved', but rather one of 'solution begun'.

8 LEARNING OUTCOMES

After you have studied Block 4 Part 1 you should be able to do the following.

1.1 Plot and interpret graphs showing variations in electricity demand over time. (SAQs 1.1 and 1.10)

1.2 Comprehend the magnitude of surges in the UK's electricity demand. (SAQs 1.1 and 1.10)

1.3 Identify different forms of energy, and transformations between them. (SAQ 1.2)

1.4 Formulate a functional specification for large-scale storage of electrical energy. (SAQ 1.6)

1.5 Understand and use the concept of conservation of energy to calculate changes in kinetic energy, gravitational potential energy and strain energy. (SAQs 1.7, 1.8 and 1.9)

1.6 Calculate energy and power losses during a process, and express these losses in terms of the efficiency of the process. (SAQs 1.3, 1.4, 1.5, 1.8 and 1.9)

1.7 Determine the energy and power capabilities of a pumped storage power station. (SAQs 1.7, 1.8 and 1.9)

1.8 Calculate the response time of a pumped storage power station. (*Mathcad* task)

ANSWERS TO EXERCISES

Exercise 1.1

(a) Table 1.10 is my answer to this question. Your figures could be quite different, particularly if you make use of cheaper night-time electricity for hot water and/or space heating.

Table 1.10 Likely electrical appliance usage in a typical household

Appliance	Typical power rating / kW	Energy consumed / kW h		
		Daytime	Evening	Night-time
TV, video, DVD, hi-fi, PC	0.2–0.5	0.5	0.5	
Vacuum cleaner	0.3	0.1		
DIY power tools	0.2–1.0	0.5		
Kitchen gadgets	0.2–3.0	0.4		
Fridge	0.3	2.4	1.2	3.6
Freezer	0.6	4.8	2.4	7.2
Washing machine	1.0–3.0	1.0	1.0	
Hot water (if electric only)	3.0	24	12	
Space heating (if electric only)	3.0	24	12	

(b) Demand is greater during the daytime and lower at night-time.

(c) Demand from manufacturing industry is relatively uniform with some factories working though the night, whereas most of the demand for electric transport will coincide with rush hours.

Exercise 1.2

$$P = \frac{E}{t} = \frac{400 \times 10^3\ \text{J}}{200\ \text{s}} = 2000\ \text{W} = 2\ \text{kW}$$

(b)

$$\frac{2800 \times 10^6\ \text{W}}{2 \times 10^3\ \text{W litre}^{-1}} = 1.4 \times 10^6\ \text{litre}$$

(c)

$$\frac{1.4 \times 10^6\ \text{litre}}{0.2\ \text{litre}} = 7 \times 10^6 = 7\ \text{million hot drinks}$$

Exercise 1.3

My completed tables are Tables 1.11 and 1.12.

Table 1.11 Mechanical energy stores

Storage mechanism	Release mechanism	Energy store
Watch spring	Balance wheel and escapement	Elastic potential
Water tower	Water shut-off valve	Gravitational potential
Compressed-air cylinder	Gas valve	Elastic potential
Flywheel	Clutch and gearbox	Rotational kinetic

Table 1.12 Electrical energy stores

Storage device	Release mechanism	Energy store
Capacitor	Switch	Electrical potential
Battery	Switch	Chemical potential
Flywheel-motor-generator	Switch or clutch and gearbox	Rotational kinetic

Exercise 1.4

(a) thermal energy

(b) chemical energy

(c) electrical energy, thermal energy and light.

Exercise 1.5

(a) The (rotational) kinetic energy used to turn the key is converted to elastic strain energy stored in a spring.

(b) The chemical energy stored in the fuel is converted to thermal energy. This in turn is converted to gravitational potential energy and (translational) kinetic energy as the balloon starts to take off.

(c) Chemical energy in the battery is converted to electrical energy, which is converted to thermal energy and optical energy. Note that the thermal energy is dissipated not just in the bulb, but also in the battery and the conductors in the torch.

(d) The translational kinetic energy of the bike and rotational kinetic energy of the turning wheels are converted to thermal energy due to friction between the brake pads and the wheels.

Exercise 1.6

You should have identified two of the following quantities as being important: power available, response time, total energy stored or else maximum duration, recovery time.

Exercise 1.7

You may have suggested any one of a number of ingenious ideas. Here are three of my suggestions:

- A large bank of batteries is charged when supplies are plentiful. When extra generating capacity is required the batteries are used to drive a motor that is coupled to a 2000 MW three-phase generator.
- A large flywheel in a strong evacuated steel casing is spun up when supplies are plentiful. When extra generating capacity is required a 2000 MW three-phase generator is coupled via a clutch to the shaft of the flywheel.
- Air is pumped into a large pressure vessel when supplies are plentiful. When extra generating capacity is required the compressed air is used to drive a turbine coupled to a 2000 MW three-phase generator.

ANSWERS TO SELF-ASSESSMENT QUESTIONS

SAQ 1.1

(a) The actual electricity demand curve during the 1999 solar eclipse is shown as the black curve in Figure 1.16.

(b) Interest in the eclipse meant that people stopped what they were doing in offices and at home to watch the event, either outside or on television.

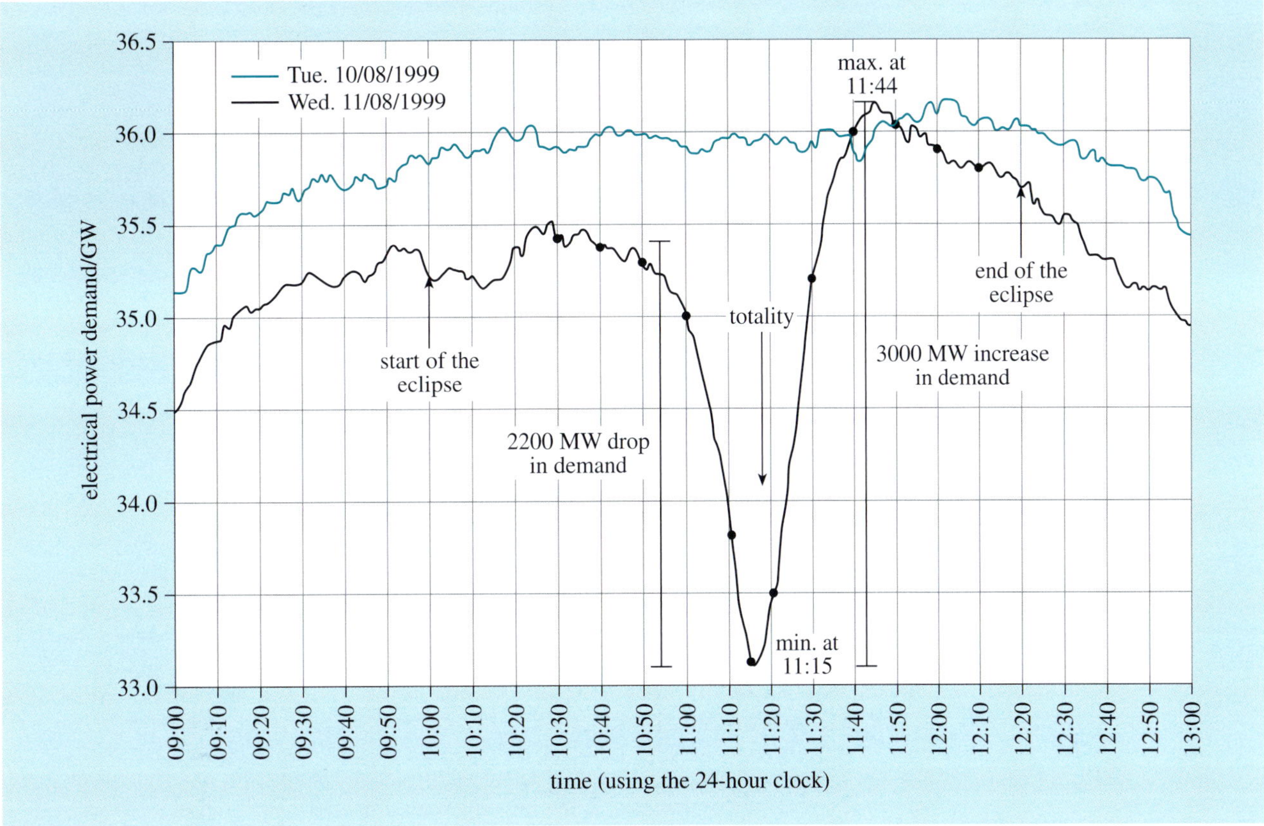

Figure 1.16

SAQ 1.2

(a) Mechanical energy, electrical energy, thermal energy, chemical energy, nuclear energy.

(b) *Mechanical energy* is the energy associated with movement, or potential for movement, of machinery and mechanisms.

Electrical energy is the energy associated with the movement, or potential for movement, of electrons.

Thermal energy is the energy associated with the movement of atoms and molecules.

Chemical energy is the energy released when bonds are formed between atoms.

Nuclear energy is the energy released when bonds are formed between protons and neutrons.

SAQ 1.3

(b) and (d) both involve heat engines.

(a) and (c) will both generate heat as a result of motion, but only as an unwanted by-product. They do not depend on a conversion between thermal or mechanical energy and work.

SAQ 1.4

(a) The efficiency is the ratio of the useful output energy to the total input energy, or the ratio of the useful output power to the total input power.

(b) $$\eta = 1 - \frac{T_2}{T_1}$$

SAQ 1.5

(a) Chemical energy available per second = 148 × 27 MJ = 3996 MJ.

Input power = 3996 MW.

(b) The operational efficiency is given by:

$$\eta = \frac{\text{useful output power}}{\text{total input power}} = \frac{800\ \text{MW}}{3996\ \text{MW}} = 0.20$$

Expressed as a percentage, the efficiency is 0.2 × 100 = 20%.

(c) $T_1 = 343 + 273 = 616$ K and $T_2 = 186 + 273 = 459$ K

Thermodynamic efficiency is given by:

$$\eta = 1 - \frac{T_2}{T_1} = 1 - \frac{459}{616} = 0.254, \text{ or } 25.4\%$$

SAQ 1.6

(a) We saw in Section 1.2 that (at the time of writing, 2003) the largest-ever TV pickup experienced in the UK was 2800 MW after the 1990 World Cup semi-final, and that the greatest-ever increase in demand was 3000 MW after the solar eclipse of 1999. The demand peak lasts for several tens of minutes. Therefore, to be useful in coping with sudden surges in demand for electricity, an electrical energy reserve must have a power output of the order of gigawatts and must be able to sustain this output for at least half an hour. My suggestions for the completed table are in Table 1.13. I've added a column of comments; your answers may differ a little.

Table 1.13 Functional specification

Generator technology	Standard three-phase	
Power / MW	2000	Peaks are of this size, but there may be more than one reserve system.
Response time / s	120	The peaks rise on this timescale.
Energy / GJ	5400	One peak has an area of about $1/2 \times 2000$ MW $\times (30 \times 60)$ s $= 1800$ GJ. Allow for three such peaks before other reserves come online.
Recovery time / h	24	TV pickups may recur daily.
Recharge technology	Electrically driven	

(b) Whatever scheme is used it will in effect be a large power station, though possibly one using 'unconventional fuel'. Among the factors that will bear on the choice of location are the following (two were asked for):

environmental impact (visual, acoustic, microclimatic, etc.)

proximity to the power grid and ease of connection to it

site access for construction, operation and maintenance

compatibility with proposed scheme (e.g. hydroelectric schemes may not be suitable in the absence of natural hills).

The specification that is emerging is effectively that of a large power station but with two distinct differences. It is unlikely to involve conventional thermal cycles owing to the need for rapid response – too much time would have to be spent just heating up huge masses of water, steel and concrete. Also, it may well not use conventional fuel – the stored resource must be replenished electrically. Another issue to bear in mind is that there will have to be some watch kept on efficiency ready for an analysis by the accountants of the operational costs and benefits.

SAQ 1.7

(a) $m = \rho V = 1000 \times 7 \times 10^6 = 7 \times 10^9$ kg

(b) The upper reservoir maximum water level is 634 m above sea level and the water level range is 34 m. So the upper reservoir minimum water level is $634 - 34 = 600$ m above sea level. The average level of the upper reservoir is $(634 + 600) / 2 = 617$ m.

(c) The lower reservoir maximum water level is 106 m above sea level and the water level range is 14 m. So the lower reservoir minimum water level is $106 - 14 = 92$ m above sea level. The average level of the lower reservoir is $(106 + 92) / 2 = 99$ m

(d) Average height

$h =$ difference in average water levels

$= 617 - 99 = 518$ m

(e) Gravitational potential energy, $U_{grav} = mgh$

$$= 7 \times 10^9 \times 9.81 \times 518$$

$$= 3.56 \times 10^{13} \text{ J}$$

$$= 35\,600 \text{ GJ}$$

(f) 7 000 000 m^3 of water empties at 390 $m^3\ s^{-1}$.

The time taken to completely empty $= 7 \times 10^6 / 390$

$$= 17\,949 \text{ seconds}$$

$$= 17\,949 / 3600 \text{ hours}$$

$$= 4.99 \text{ hours, or approx. 5 hours}$$

(g) EITHER

power output = available energy / time

$$= 3.56 \times 10^{13} \text{ J} / 17\,949 \text{ s}$$

$$= 1.98 \times 10^9 \text{ W} = 1980 \text{ MW}$$

OR

volume flow rate × density = mass flow rate

power = potential energy usage rate = mass flow rate × (gh)

$$= 390 \text{ m}^3\ \text{s}^{-1} \times 1000 \text{ kg m}^{-3} \times (9.81 \text{ m s}^{-2} \times 518 \text{ m})$$

$$= 1.98 \times 10^9 \text{ W} = 1980 \text{ MW}$$

SAQ 1.8

(a) Total capacity $= 6 \times 288$ MW $= 1728$ MW

(b) The predicted power available from the gravitational potential energy of the water is greater than the actual power output of Dinorwig power station.

(c) (i) $\eta = 1728 / 1980 = 0.87$

(ii) $\eta = 1728 / 1980 \times 100\% = 0.87 \times 100\% = 87\%$

SAQ 1.9

(a) From SAQ 1.7 part (a), total input energy = 35 600 GJ

useful output energy = total input energy × efficiency

$$= 35\,600 \times 0.87 = 30\,972 \text{ GJ}$$

potential energy lost = total input energy − useful output energy

$$= 35\,600 - 30\,972 \text{ J} = 4628 \text{ GJ}$$

(b) Energy losses occur in the following situations, all of which arise in hydroelectric systems:

friction at the walls of the pipes (changing mechanical energy to heat)

residual kinetic energy not extracted from the water after passing through the turbines (some kinetic energy transferred to the lower pond) generation of sound as the water rushes through the tunnels and turbines

the electrical-to-electrical conversion in transformers and currents in conductors

friction in bearings.

SAQ 1.10

(a) See Figure 1.17.

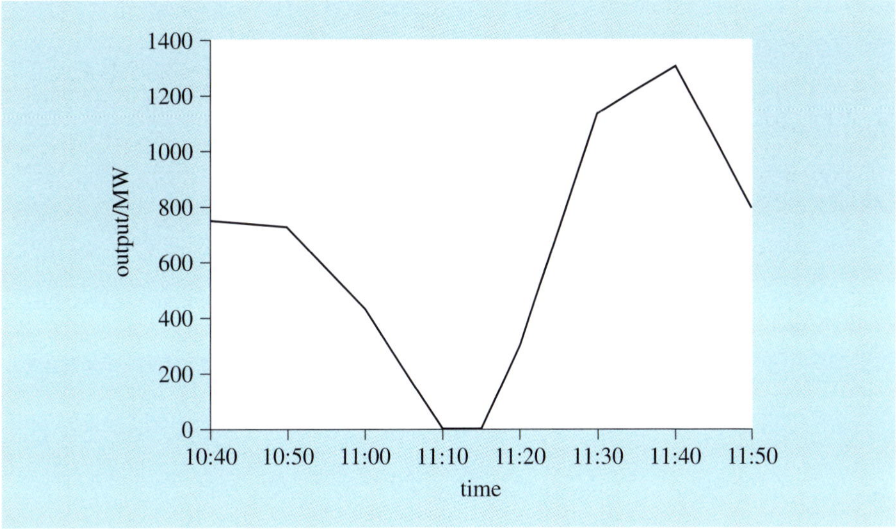

Figure 1.17

(b) Between 11:15 and 11:40, the output of Dinorwig increased from 0 MW to 1300 MW. So, Dinorwig contributed about 1300 MW to cope with the 3000 MW surge in demand. Dinorwig provided 1300 / 3000 = 43% of the extra electricity required. Equally important is the fact that Dinorwig was already running and could be quickly closed down at the start of the eclipse, as demand dropped.

ACKNOWLEDGEMENTS

Grateful acknowledgement is made to the following sources for permission to reproduce material within Part 1 of this block.

Figures

Figures 1.2 and 1.3: Extracted from www.4seasons.org.uk, © The Federation of Groundwork Trusts and cmn design; *Figures 1.12 and 1.13:* Courtesy of Dinorwig Power Station.

Course team acknowledgements

Part 1 of this block was prepared for the course team by Nick Braithwaite, David Sharp and Jan Kowal.

Trademarks

Mathcad® is a registered trademark of MathSoft Engineering & Education Inc., USA.

Part 2
The Francis water turbine

CONTENTS

1 INTRODUCTION

1.1 Aims

The overall aims of Part 2 of this block are:

- To illustrate the many ways in which technological development of an idea can be progressed.
- To show, through the example of the history of the development of the water turbine, how society's needs drive and direct engineering effort.
- To apply the ideas of the conservation of energy and momentum introduced in Block 3 to the design of two different types of water turbine.

1.2 Introduction to the Francis water turbine

> *The American mixed-flow turbine, now commonly referred to as the 'Francis' turbine ... was not a sudden invention; it was merely the crystallization and modification of principles towards which all had been working. Many an old, self-trained mechanic contributed his mite to the development of this new type. This was a real American production, the result of evolution during a changing period in American history. The need arose, made itself felt, and eventually was met, not by the work of one great scientist, but by the multitudinous efforts of an army of old Yankee millwrights and machinists, many of the names of whom are either unknown or forgotten.*
>
> (Arthur Safford and Edward Hamilton, 1922)

The quotation you've just read is a description of one way in which a device evolves over time until it's as good as the available materials and manufacturing methods allow it to be. So gradual is the process that sometimes it is hard to identify when it was invented and by whom. A large chunk of this Part will take the Francis water turbine and examine its origins in detail. This historical account focuses on how progress from stage to stage was driven and directed by needs. It also shows how different kinds of innovative steps were each appropriate at different times, according to the nature of those needs and the cultural and scientific context.

The Francis turbine, named after James Bicheno Francis (1815–92), one of America's most distinguished nineteenth-century hydraulic engineers, is today one of the most versatile and widely used of all water turbines. It is suitable for operation at both high and low head (height difference between the reservoir feeding the turbine and its exit tube), and is available 'off the shelf' in sizes ranging from a few kilowatts to several hundred megawatts.

Water turbines are sometimes referred to as 'hydraulic motors'.

Its construction and operation is discussed qualitatively in **▼The Francis water turbine▲** and more mathematically in Section 5.

One of my main aims is to examine how novel ideas are turned into successful innovations. In particular, the study explores two distinct but interrelated techniques by which inventions are modified and improved so that they are better matched to the needs of the market. The first approach (of which Francis himself was a prime exponent) involves the use of theoretical knowledge as a guide to product development – what is sometimes referred to as 'rational design'. The second involves a more pragmatic experimental approach, which relies to a greater extent on intuition and craft skills. As the study demonstrates, both approaches played their part in the conception and evolution of the Francis turbine.

▼The Francis water turbine▲

The Francis turbine contains a compact wheel, or runner, which is mounted either horizontally (as in Figure 2.1) or vertically. It is connected directly to the main drive shaft and operates under the pressure of a head of water.

The precise shape of runner can vary a great deal depending on the conditions of head and flow under which it operates and the speed at which it is designed to turn. It is cast in one piece and contains a number of fixed, spoon-shaped blades, which channel water first inwards towards its axis, then downwards along its axis before it is discharged into the tail race. This flow pattern is what hydraulic engineers refer to as 'mixed flow'. The water enters radially and is accelerated into the runner by gradually restricting the passages between the blades. It is then turned through a smooth 90° bend, decelerated (by increasing the cross-section) and discharged in a predominantly axial direction. All the time, its pressure falls as it passes down between the blades until, on discharge, it is close to atmospheric pressure.

The Francis turbine is designed to operate fully submerged; to ensure that it is always full of water, a flared tube (called a draft tube) is fitted on the discharge end of the turbine, with its far end below the water level of the tail race. The flared shape of the tube ensures that the water coming out of the turbine is quickly taken out of the way.

The Francis turbine is a 'reaction' turbine. Its runner spins as a result of both the pressure of the water reacting on its blades and the impulse produced as the water's speed and direction change.

In most types of Francis turbine, the runner is mounted within a spiral casing known as a scroll. This forces the water to flow in a circle around the runner. Within the scroll, two rings of guide vanes – one fixed, the other moveable – impart an inward-going component to this motion. This ensures that the water enters the spinning runner at a speed matched to that of its outer edge.

This arrangement of scroll case and guide vanes also guarantees that there is full admission of water and that the runner remains submerged at all times.

The inner guide vanes are also used as a means of controlling the output from the turbine; they are mounted on pivots and can be adjusted to increase or decrease the flow onto the runner in accordance with the demand for power. Operation of the guide vanes is controlled by a governor which senses when the output from the runner is out of balance with the demand and takes corrective action automatically.

Many of the elements of the turbine are visible in Figure 2.2.

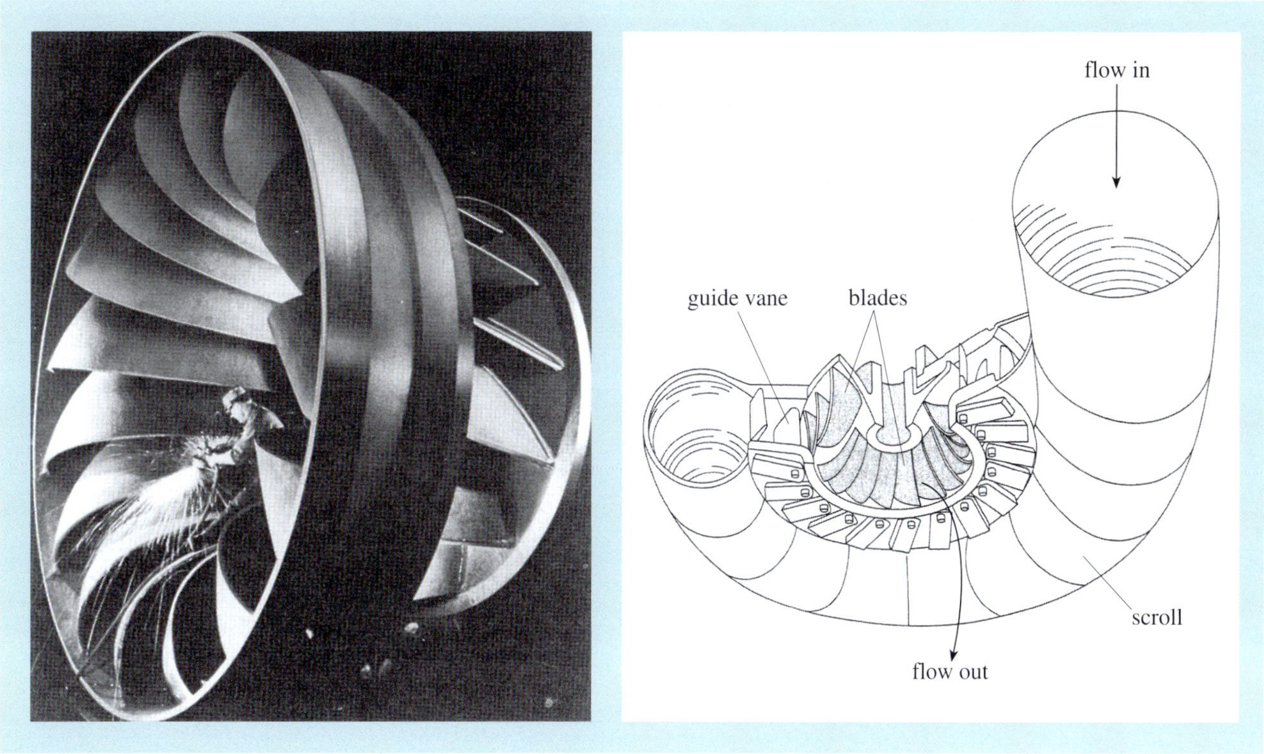

Figure 2.1 Francis turbine

Figure 2.2 Francis turbine workshop. The man in the white coat is inspecting a runner for a Francis turbine manufactured by Gilkes in Kendal, England. Scroll casings and a guide vane control mechanism can also be seen

James Francis was born in England in 1815, but emigrated to the United States of America after serving a brief engineering apprenticeship on the design and construction of harbours and canals in the south of England. He was only 18 at the time.

Within a year of his arrival in the USA, he had secured work with the Proprietors of Locks and Canals at Lowell, the prosperous textile-manufacturing centre in Massachusetts. It was not long before he demonstrated his considerable practical and managerial abilities. He assumed the job of Chief Engineer after only three years with the company and in 1844 took on additional responsibilities for supervising the smooth operation of the entire water system that supplied power to the mills. He held the position of Chief Engineer and Superintendent of Locks and Canals for over 40 years, and maintained his links with the company right up to his death in 1892.

It was at Lowell that Francis carried out fundamental studies on water turbine design, work that ultimately led to the development of the turbine that now bears his name. I shall examine Francis's role in this important development in due course. But first I should like to look at the early history of water power and to focus, in particular, on the evolution of large industrial water wheels – the predominant source of power at Lowell in the 1840s – and the conception of the reaction water turbine. The first prototypes of this new class of prime mover were produced in France a quarter of a century before Francis began his experiments.

This history reveals that many of the ideas that form the basis of modern hydraulic engineering evolved out of practical and theoretical investigations into water wheel design. It is of special interest in the study of innovation since it not only involves the evolution of scale modelling techniques and controlled testing, which are today used as a matter of course in product design and development, but is also one of the earliest applications of theoretical knowledge to the development of improved products.

2 THE RISE AND FALL OF THE WATER WHEEL

The predominant form of water wheel in use at Lowell and other industrial centres in the mid-nineteenth century was the breast wheel. This was a cross between two much older water wheels which operated on entirely different principles. The first was the undershot wheel, which evolved around 2000 years ago and was one of the earliest prime movers. The second was the overshot wheel which appeared on the scene several centuries later.

The first reliable and unambiguous reference to the use of undershot water wheels is to be found in the writings of the Roman engineer Marcus Vitruvius around AD 25. Vitruvius describes wheels with 'float boards' fixed around their faces which are turned by 'the mere impulse of the river, without any treading on the part of workmen'.

At that time, the wheels were used primarily for milling grain (see Figure 2.3). Their axles were extended to form a drive shaft connected, via wooden cogs and gears, to a millstone fed with grain from a hopper.

Figure 2.3 A traditional undershot water wheel used for milling grain

It would be difficult to conceive of a more simple arrangement for a prime mover: there was only one moving part, the wheel itself, which was turned simply by the impulse from running water.

What is more, undershot wheels could be built relatively cheaply from locally available materials using the crudest of tools and the minimum of craft skills, and could operate almost anywhere that flowing water was to be found. They could even be mounted on river barges anchored midstream and turned by the current. It was this flexibility that explains the almost universal appeal of the undershot wheel.

Undershot wheels were custom built to suit specific power requirements and site conditions. The speed of rotation of the wheel depended on both the speed of water striking the floats and the wheel's diameter. The smaller the wheel, the faster it turned. Power output was determined by the quantity of water passing under the wheel in a given time. The most powerful wooden wheels built to this simple formula probably generated no more than 7.5 kilowatts.

By the late Middle Ages, the range of applications of the Vitruvius water wheel had been greatly extended to include paper-making, fulling cloth, boring and sawing wood, crushing ores, pumping flood water from mines, lifting materials and working metals. Indeed some medieval developments, notably metalworking, would not have been possible without the concentration of power that water provided.

Thus water power played a vital role in the medieval power revolution, just as it did some centuries later in the Industrial Revolution. It was, furthermore, an important stimulus to the development of fundamentally important mechanisms such as the tappet, crank and connecting rod, which converted rotary motions into reciprocating actions – either back and forth or up and down. Reciprocating drives were needed for many industrial processes, such as sawing and working forge hammers and bellows.

It is somewhat surprising that many centuries of use saw very little modification to the basic design of the undershot wheel, despite some rather obvious limitations, such as its low energy-conversion efficiency, short operational life and susceptibility to breakdown. This was largely a consequence of the conservatism built into the millwrighting profession: millwrights were basically craftsmen and not designers. They applied age-old rules of thumb when determining the correct size and proportions of a wheel, and were understandably reluctant to venture far from proven traditional patterns of design and construction.

The situation did not begin to change until the sixteenth century, when the first practical manuals on water power began to appear. These early handbooks contained a wealth of practical advice and tables which related the size and proportion of the wheel to its output under different combinations of head and flow rate.

The sixteenth century also saw a major challenge to the supremacy of the ubiquitous undershot wheel. The overshot water wheel was turned not by the impulse of running water, but by its weight. Floats were replaced by buckets which were filled from an overhead sluice as the wheel revolved (see Figures 2.4 and 2.5)

Overshot or 'gravity' wheels required more massive constructions than undershot wheels because of the extra weight of the buckets and the water they carried. They were more difficult to build and consequently more expensive. However, they were widely considered to be more efficient and were therefore favoured over the undershot wheel wherever there was sufficient head of water to permit their use.

Quite how far the performance of the overshot wheel exceeded that of the undershot Vitruvian wheel was not fully appreciated until the middle of the eighteenth century when a series of carefully conducted experiments was performed in England and France. These experiments had far-reaching effects. Not only did they influence the whole course of development of the water wheel, and subsequently the water turbine, but they also had a fundamental impact on the way technical problems were approached and resolved.

Figure 2.4 Sixteenth-century drawing by Agricola of an overshot water wheel with a rag-and-chain pump

Figure 2.5 Sixteenth-century drawing by Agricola of an overshot water wheel with a reversing double-wheel hoist

2.1 The beginning of engineering design

Two of the key figures in the evolution of this new experimental technique were John Smeaton and Antoine de Parcieux.

Smeaton was an Englishman of considerable mechanical talent who is generally regarded as being one of the first civil engineers. He is perhaps best known for his work on the building of marine installations in the south of England, including the Eddystone Lighthouse.

De Parcieux was a gifted French mathematician and something of an expert on water supply. It was he who was called upon by the Paris Académie des Sciences to investigate a hydraulic problem posed by Mme de Pompadour, Louis XV's celebrated mistress, who wanted a secure supply of water at her château at Crécy.

The contribution that Smeaton and de Parcieux made to the methodology of engineering design was considerable. Working independently in their respective countries they pioneered the use of two powerful techniques – parameter variation and scale model analysis – which are today used routinely by engineers and product designers for perfecting innovations.

2.1.1 Parameter variation

Parameter variation is a method for optimizing the design and performance of a process or product. It involves first identifying the factors that affect the behaviour of the system, and then carrying out a systematic series of experiments in which each parameter is changed in turn. In the case of water wheels, the critical parameters include such factors as the speed of rotation of the wheel, the size and shape of the floats or buckets, the head of water and its rate of flow, and the load on the wheel. Careful interpretation of experimental results can provide useful information about the relationship between the parameters themselves and the effect that a change in any one can have on the operation of the system as a whole. Such data can provide guidance as to how a process or product can be improved.

Smeaton successfully applied parameter variation to the development of a marine cement that would set firm under water, and to improving the Newcomen steam engine – his 'rationalized' version was twice as economical on fuel as the original.

Exercise 2.1

Take a few minutes to think about parameter variation as a means of optimizing a device or process. What limitations can you see for the method?

2.1.2 Scale model analysis

Another tool of the designer is modelling. It is a means of evaluating the behaviour of full-sized systems without going to the trouble and expense – and often the risk – of building them. The techniques developed by Smeaton and de Parcieux involved building and testing scale models of different designs of water wheel and measuring their performance. By interpreting the results of the tests, they were able to make accurate predictions about the performance of industrial-sized wheels. Smeaton himself considered 'experiments on working models … as the best means of obtaining the outlines in mechanical enquiries'.

Today, scale model analysis is only one of a number of modelling techniques that are used routinely by technologists. Other methods include, for example, graphical, mathematical and computer modelling. However, scale models remain an extremely valuable tool for use in situations where other techniques cannot easily be applied, for example where a process or behaviour pattern is too complex to be satisfactorily described using equations or algorithms.

2.2 The experiments of Smeaton and de Parcieux

Smeaton began his experiments in the early 1750s. He built precision models of both undershot and overshot water wheels, and tested these on a special rig (see Figure 2.6) to see how they performed under different operating conditions.

To test undershot wheels he used a rig with an open channel which was supplied with a continuous flow of water from an elevated tank. The water flowed through a sluice and along the bed of a flume before being

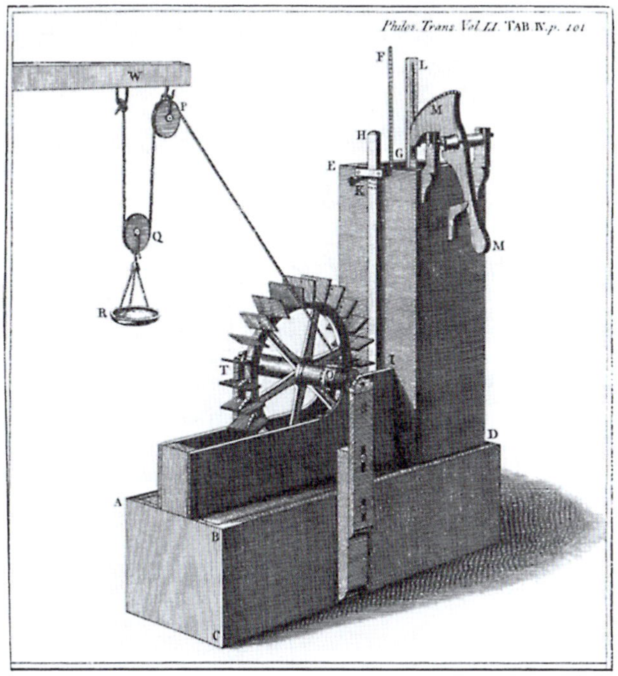

Figure 2.6 Smeaton's experimental test rig

discharged into a sump, from where it was pumped back into the header tank using a hand pump.

He controlled the flow of water by adjusting the sluice and was able to calculate the rate of flow from the number of pump strokes required over a given time to maintain a constant head in the tank.

A scale pan was connected to the axle of the wheel via a rope and pulley system. Thus Smeaton could measure the power output of the wheel by recording how fast it would lift a given set of weights through a given distance. (We did something similar to assess the performance of electric motors in Block 3 Part 3.) He also adapted his apparatus for overhead delivery of the water and carried out a similar set of tests on model overshot wheels.

There was much interest in milling circles when Smeaton's results were published by the Royal Society in 1759, since they resolved a smouldering controversy as to whether undershot or overshot wheels were more efficient. Some prominent individuals of the early eighteenth century had argued that undershot wheels were as much as six times more efficient than overshot wheels. Others took the contrary view, arguing that weight had a clear advantage over impulse as a motive power in water wheels.

Smeaton's results were conclusive. He found that a good overshot wheel could be more than twice as efficient as an undershot wheel. From that point on, Smeaton, who was an influential figure in his day, recommended the use of overshot wheels wherever possible and ▼**Breast wheels**▲ where insufficient head was available.

Breast wheels became increasingly important in the late eighteenth century, and rapidly displaced the simple undershot wheel. They were well suited to the operating requirements of large mills and factories and offered several major advantages over the traditional overshot wheel. Smeaton's basic designs of overshot and breast wheel were widely copied in Britain and abroad and helped to sustain the role of water power against competition from the steam engine until well into the nineteenth century.

▼Breast wheels▲

The breast wheels of John Smeaton were essentially undershot wheels with a large number of flat radial blades – often as many as 40 – but they operated on a combination of impulse and gravity. Beneath the wheel was a close fitting apron or 'breast' made from wood or stone (see Figure 2.7). This held the water onto the wheel as it fell, only allowing it to escape into the tail race at the very bottom of the wheel's turn when its gravitational energy had been largely expended.

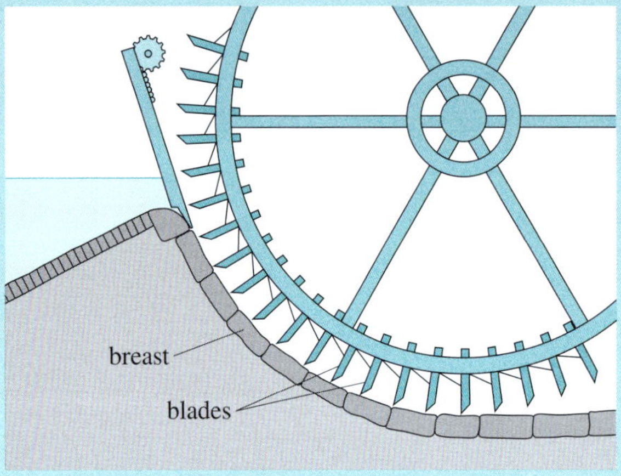

Figure 2.7 Drawing of Smeaton's recommended breast wheel

The size of the breast varied according to the head of water available. As the head increased, the floats tended to get replaced by buckets, so that high breast wheels came close to becoming overshot wheels: they were constructed along similar lines. High breast wheels presented four important advantages over the overshot wheel:

- They could be built with a diameter considerably greater than the available head. This offered greater operating flexibility in adjusting the speed of revolution to the working requirements.
- The fact that water was delivered to the wheel off-centre reduced the weight on the axle, permitting lighter construction. (Water in buckets immediately above the axle, as in the traditional overshot design, does not exert a turning force on the wheel.)
- The height of the sluice could be changed in response to seasonal or short-term fluctuations in water level.
- Breast wheels were less susceptible to loss of power as a result of backwater flooding because the lower part of the wheel rotated with, rather than against, the direction of flow of water in the tailrace.

De Parcieux carried out his experiments at about the same time as Smeaton and, not surprisingly, arrived at similar conclusions. But he extended the work to examine the effect on the performance of the undershot wheel of varying the angle of the blades (see Figure 2.8). He determined, through experiment, that if the blades were fixed at 30° to the normal radial setting, then the wheel's efficiency doubled because the blades entered the water with less splash and turbulence (which are effectively a waste of energy).

De Parcieux had taken an important step towards 'shockless entry', an essential condition for efficient hydraulic motors.

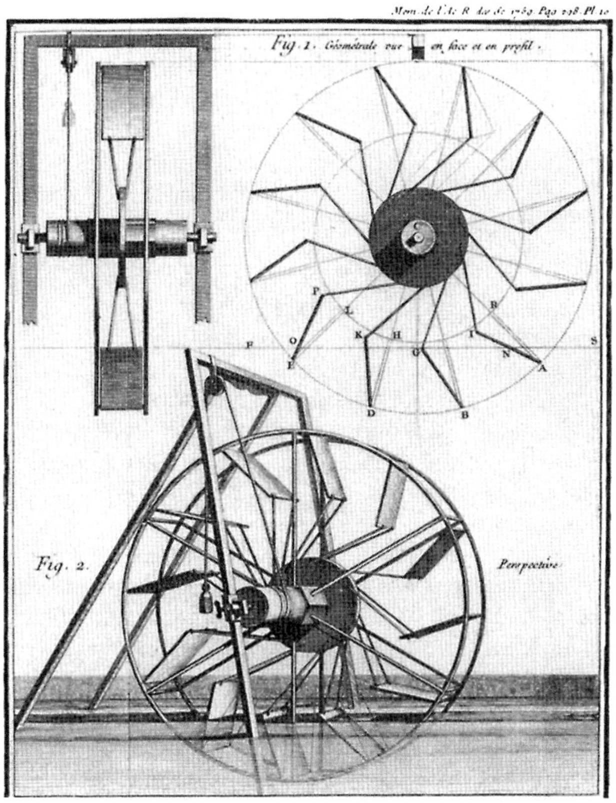

Figure 2.8 De Parcieux's apparatus for testing the comparative effect of varying the angle of the blades, 1759

Smeaton and de Parcieux were not the first to use models as an aid to understanding or evaluating technology. However, they were among the first to focus on quantitative concepts and to test complete machines in accordance with clearly defined principles.

The understanding of the basic mechanics of water wheel design evolved rapidly as a result of Smeaton and de Parcieux's careful empirical studies. Others followed, extending and developing the work, and investigating, for example, the relationship between the output of the wheel and the speed at which it turned in streams of different flow speed; the effect of increasing the number of blades and changing their size in relation to the wheel's diameter; and the optimum depth to which the blades of an undershot wheel should be submerged.

Important though it was, this experimental work could go only so far. It clarified problems and provided greater understanding of mechanisms but did not of itself lead to new inventions; rather it was a method of perfecting known inventions by incremental development.

New insight was needed to break out of the straitjacket of conventional thinking, and this was provided by advances in theoretical understanding. It was these that paved the way for the modern water turbine.

2.3 The development of a theoretical approach to design

One of the first individuals to investigate water power from a theoretical perspective was Leonardo da Vinci in the late fifteenth century. Da Vinci appreciated that water's ability to perform work depended on the distance it was free to fall, and that not all of its potential could be realized as useful mechanical energy because of unavoidable frictional losses in the prime mover and its mechanical linkages (see Figure 2.9). He was also the first person to recognize that undershot wheels and overshot wheels worked on different principles, one responding to the water's impulse, the other to its weight.

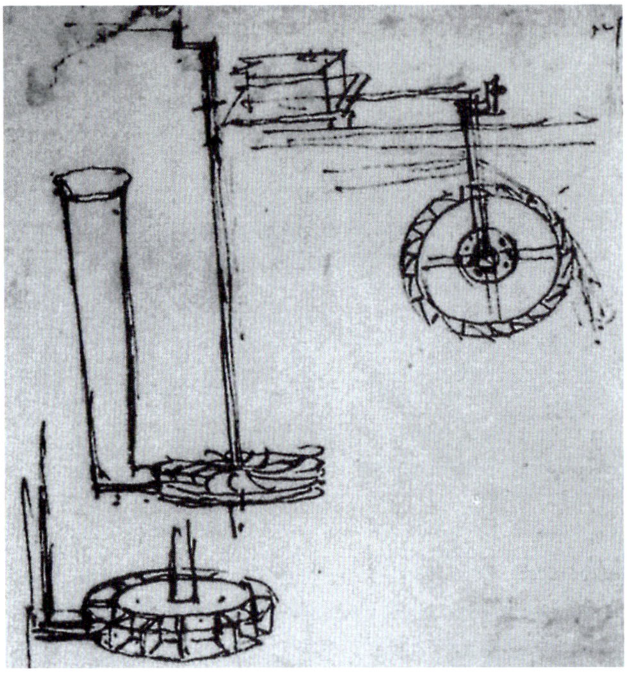

Figure 2.9 Three water wheels activating a reciprocating motion, from the *Codex Atlanticus* of Leonardo da Vinci at the Ambrosia Library in Milan

This led to the idea that impulse and gravity wheels should be equally effective as prime movers, and this view was later reported by such notables as Galileo Galilei and Evangelista Torricelli in the sixteenth and early seventeenth centuries. Indeed, Torricelli showed in experiments that water spouting from an orifice under a given head reached the same speed as water which fell freely from an equivalent height.

Torricelli's results will be shown in Section 5.1 to fit a simple theoretical model.

There were attempts in the late seventeenth century to measure the force exerted by flowing water on a stationary plate by means of a counter-balanced weight, and to estimate from this the maximum force exerted on a wheel. However, the experiments were crude and of limited use. What was important was not the force that a stationary wheel could exert, but the *work* that a revolving wheel could deliver under different conditions of flow and head.

Despite their limited application, these rather naive experiments did lead to an investigation of the relationship between the speed of revolution of the undershot wheel and the flow speed of the stream; and this, in turn, revealed the point at which maximum power output was achieved.

The pace quickened in the eighteenth century, largely as a result of the efforts of a handful of gifted mathematicians, physicists and astronomers centred in Europe. By the end of the century there had evolved a better appreciation of the key factors which affected the performance of water wheels. There was recognition that energy was lost when water flowed along open flumes, or poured into the tailrace, and when it struck the blades of a revolving wheel, producing whirls and eddies.

Exercise 2.2

Briefly justify the statement above that 'energy was lost when water flowed along open flumes, or poured into the tail race, and when it struck the blades of a revolving wheel, producing whirls and eddies'.

The basis for a mathematical analysis of the problems had also been laid. The three individuals who take much of the credit for this are Daniel Bernoulli, generally thought of as the father of theoretical hydrodynamics, Leonhard Euler, the pre-eminent mathematician, and his son Johann.

Bernoulli calculated the theoretical maximum efficiency of undershot wheels. His *Hydrodynamica*, published in 1738, was a classic in its time. His work was later used by Johann Euler to carry out a theoretical investigation of the merits of overshot and undershot wheels from which the clear superiority of the overshot wheel was established. This work was concurrent with that of Smeaton and de Parcieux but not known to them.

Leonhard Euler also worked on vertical water wheels, but his main contributions were to the understanding of early reaction wheels, devices like modern lawn sprinklers, which are propelled around by the reaction from a jet of water. He developed the first comprehensive mathematical theory of the turbine.

But the individual who drew much of this work together and first formulated the fundamental requirements for the perfect hydraulic prime mover was Jean Charles Borda. Borda's thesis, contained in a paper submitted to the Paris Académie des Sciences in 1767, has guided the designers of water wheels and water turbines ever since. It was Borda's contention that, for maximum conversion of energy to occur, the water should enter the motor without shock and leave without speed.

However, despite the insight that Borda brought to the field of water power, it was almost 50 years before his principles were successfully applied to the development of the undershot wheel and its natural descendant, the water turbine. In the meantime, other revolutionary changes were taking place which were to have an equally dramatic effect on the design and construction of hydraulic motors.

2.4 The impact of new materials

The introduction of new materials, most notably cast, and later wrought, iron, radically changed the way wheels were made and how they performed. The introduction of power take-off from the rim of the wheel rather than the axle was also of fundamental importance.

Iron not only gave greater strength and durability to water wheels, but it also made possible greater precision in design and construction and greater smoothness and efficiency in operation. Fixing the main driving gear to the outer perimeter of the wheel meant that the main shaft was no longer needed to transmit the power: it merely supported the weight of the wheel. The massive wooden axles and heavy spokes that had been necessary in large wheels to cope with the very considerable torques disappeared.

Wheels of more elegant design appeared. They had slender axles and wire spokes, and were capable of far higher rotational speeds and increased power output. The largest wheels delivered several hundred kilowatts (see Figure 2.10).

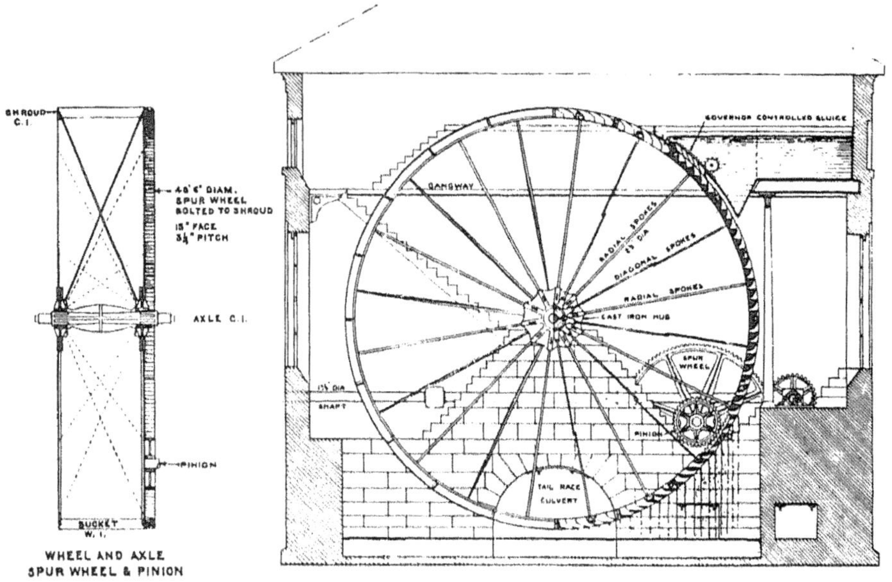

Figure 2.10 The pitch back wheel: a modified form of overshot wheel similar to the high breast wheel, which was designed and manufactured by Sir William Fairbairn in 1827 and built at the works of James Finlay and Co. Ltd, Catrine, Ayr. Two wheels were built. They had cast-iron shafts with wrought-iron radial and diagonal spokes which were kept in tension. Each wheel was over 15 metres in diameter and 3 metres wide, and could generate 190 kilowatts

John Smeaton had been one of the first to recognize the potential for using iron in water wheels and had experimented with cast-iron shafts and other components. However, the construction of wheels made entirely of iron had been beyond the technology of his time.

Wood had several indisputable advantages over iron as a structural material. It was cheap, readily available and much easier to work, and craftsmen skilled in cutting, shaping and joining it were easily found. But its limitations and deficiencies were also very apparent, especially to the mill owners. Historian Louis Hunter in his study of water power in the USA notes that:

> *Against wood's advantages as a structural and mechanical material must be placed its deficiencies, especially in the moving parts of machinery and as compared with iron: a lack of strength; a readiness to wear, decay, warp, and shrink, and, under conditions of intermittent use and alternate wetting and drying, to loosen at the joints. The large dimensions necessary to offset the low strength of even the best woods – ranging from one-fourth to one-tenth the strength of wrought iron, according to the character of the stress – required the use of shafts and gearing that were heavy, cumbersome, and power wasting in their friction. The creaking and rumbling of moving parts that in time became loose and ill-fitting announced the presence of a water mill for hundreds*

of yards about on a calm day. Wheel shafts two feet [60 cm] in diameter were common, and in the larger mills the shafts reached three feet [1 metre] and more. An overshot wheel of eighteen foot [5.5 metre] diameter, with main gearing to drive a two-run flour and grist mill, probably developing at most 15 horsepower [11 kilowatts], weighed when dry about eight tons [8 tonnes]. That meant a weight of about one-half ton per horsepower [two-thirds of a tonne per kilowatt], not including the weight of the water absorbed by the wood and carried in the buckets when the wheel was in operation.

Note that 'one-half ton' is 'half a ton'.

The power loss by friction in wheel and gearing may well have totalled a fourth and more of that generated. With the double gearing usual in two-run mills – consisting in millwright's terminology of master cogwheel, counter-cogwheels, wallowers and trundles – wear and tear was heavy. In busy mills the almost constant attention required to detect worn and broken cogs was a source of trouble and expense.

(Hunter, 1979)

Wooden water wheels rarely lasted more than ten years. Iron wheels, by contrast, would last three times as long, although they did require more capital outlay.

The switch from wood to iron had other consequences which went further than wheel design. Indeed, they affected the very nature of commercial exchange. Hunter points out that the use of iron not only required cash outlay by the buyer, but also:

... commercialized production, taking the fabrication of waterwheels out of the hands of local millwrights and carpenters and transferring it to the increasingly numerous and widely distributed blast furnaces and foundries by which the needs of a preindustrial economy for millwork and simple machines were largely met. In the larger ironworking establishments made possible by improved transportation and marketing facilities, shop production came to be supported by sales promotion through advertising, testimonials, and the like. The intricacy and endless modification of design possible with skilful patternmaking and encouraged by the growing demand for improved waterwheels supplied the almost wholly empirical basis for advances in wheel design and performance in the years ahead.

(Hunter, 1979)

Thus the way was open for the design and construction of the large industrial water wheels that played such an important role in the industrialization of Western Europe and North America in the early years of the nineteenth century.

2.5 The rationally designed undershot wheel

One of the first individuals to successfully apply the emerging hydraulic theories of Borda and others to the new iron water wheels was Jean Victor Poncelet, a distinguished French military engineer. Poncelet's starting point was the traditional undershot wheel which, although highly inefficient in operation, possessed a number of attractive features. Its simplicity in design meant that it could be built relatively easily and quickly, and therefore cheaply. It was, too, a robust and reliable device, at least by comparison with other water wheels of the time, and could be designed to operate across a range of heads from less than a metre to five metres and more.

The very fact that the undershot water wheel was inefficient presented Poncelet with an ideal opportunity to test Borda's principle that water should enter the wheel without shock and leave without speed. His objective was to design a wheel that combined high efficiency with low cost.

Poncelet achieved his objectives in 1823, and did so with remarkably few changes to the basic design, by using curved blades and a special sluice (see Figure 2.11).

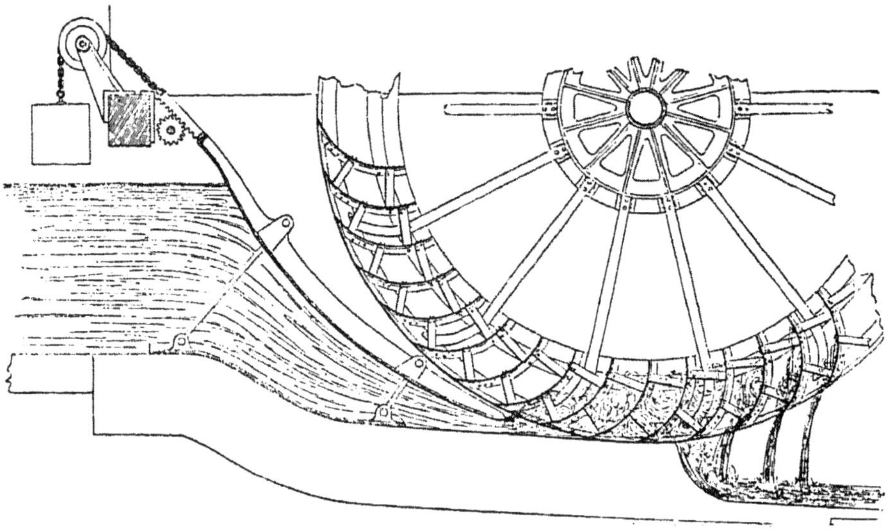

Figure 2.11 The Poncelet wheel

The changes are described by Norman Smith in the following way:

> *At the delivery site of the wheel he used a deep angled sluice to generate a broad jet whose direction coincided exactly with the leading edge of each curved blade as the latter moved down into a position to be struck by the jet. In this way, in all practical terms, shockless impingement of water on blade was achieved. As it 'climbed' the curved blade the water gradually gave up its energy. Running back along the blade would restore the water's velocity (always neglecting friction effects) relative to the blade. At exit from the wheel, however, the water's* absolute *speed is reduced to zero. Poncelet was thus able to realize, at least on paper, the theoretical ideal.*
>
> (Smith, 1976)

Poncelet's theoretical approach to design was backed up by experiments on working models and on a full-sized prototype. He determined experimentally that efficiencies of 60–70% were possible at low heads, provided the correct ratio of wheel-to-water speed could be achieved. Construction of the curved blades was the only detail of the wheel which presented any difficulties since their shape and positioning were critical. However, the blades could be readily fabricated from wrought iron sheet, which was beginning to be produced at the foundries.

Poncelet's work brought him great acclaim in France, where water power was of considerable economic importance. He was awarded the Prix de Mécanique of the Académie Royale des Sciences in recognition of his contribution to the national good.

But despite his achievements, Poncelet's 'rationally designed' undershot wheel was still far from ideal, as he himself recognized. It had two basically undesirable features. First was the fact that water entered and left the wheel at the same point (the outside edge), which meant that its flow had to reverse in the buckets and this involved energy loss. Second, the water impacted the wheel at only one position rather than simultaneously from *all* points.

Exercise 2.3

Think about the two challenges that Poncelet recognized.

(a) How could you arrange for water to enter and leave by different paths?

(b) How could you arrange for water to impact on all parts of the wheel at once?

In an important lecture of 1826, Poncelet proposed a radically new configuration that would overcome this difficulty. He suggested that the wheel might be mounted horizontally on a vertical axis and that water be introduced at all points around the wheel so that it flowed over the blades and was discharged through the centre of the wheel with the minimum of speed (see Figure 2.12).

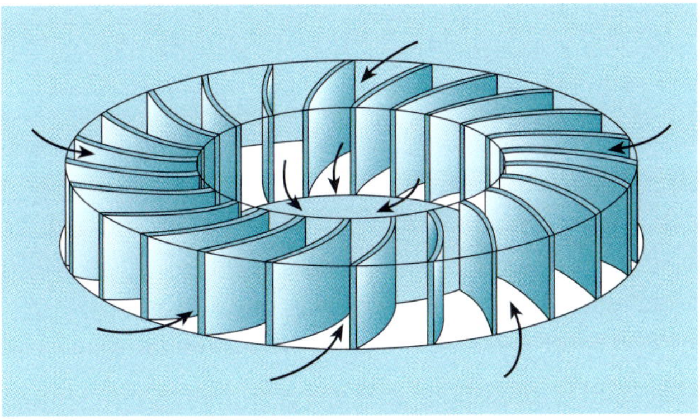

Figure 2.12 Poncelet's 'radial inward flow' proposal for a hydraulic motor

Thus Poncelet was the first to propose the radial inward-flow principle for a hydraulic motor. It was this concept which was so brilliantly applied by James Francis and others in the USA twenty and more years later.

2.6 The writing on the wall

But no level of improvement in basic design could overcome some of the most fundamental problems of water wheel operation. The concept of turning a prime mover by the action of water splashing off radial blades or falling into buckets had been developed almost to the limit. Technical improvements could no longer keep pace with the increasing demand for power or the growing requirements of factory owners for much greater security of power supply.

In an industrial setting, the limitations of the water wheel were obvious.

- It had to be very large to generate sufficient power for operating machinery. This led to considerable expense not only in fabrication, but also in providing the necessary support structures and excavating the wheel pit.
- The fact that the wheel was large meant that it turned relatively slowly. But the industrial machinery that was coming into use more often than not required a high-speed drive shaft and substantial gearing was therefore necessary. This not only added to the expense, but also proved to be very troublesome and was a major cause of stoppages. Friction in the gear train was also a source of significant power loss.
- Water wheels suffered from interruptions as a result of low water in summer and ice and flooding in winter. Flooding raised the level of water in the tail race and interfered with the wheel by exerting a braking action. Suspension of operations caused minor inconveniences to the country miller, but for the owner of a large mill it was an extremely serious problem since it meant workers, and expensive equipment, would be idle.

In short, there was a need for a smaller hydraulic motor that would be cheaper to build and install, and which would operate reliably over a wide range of heads; a motor that would be capable of much higher rotational speeds (to reduce the gearing needs) and greater efficiencies (to economize on water use), and which could operate submerged (to eliminate backwater problems). Such a motor was not long in coming. In fact it arrived in France not four years after Poncelet had announced his new wheel. The man who made the breakthrough was Benoît Fourneyron, a truly remarkable character who Poncelet once fittingly described as 'the James Watt of France'.

2.7 The first water turbine

Benoît Fourneyron was born in Saint-Étienne in 1802 of well-to-do parents descended from landed farming families. His father was a prominent land surveyor. At the age of nine, he entered the local school and soon showed himself to be an exceptionally gifted child. He was an 'all-rounder', but had a particular fondness for mathematics.

Saint-Étienne was in a thriving industrial region of the Loire and from an early age Fourneyron was encouraged to take an interest in industrial matters. His father had involved him in his surveying work, which is where he first demonstrated an extraordinary facility with mathematics. At the age of fifteen, Fourneyron entered the local School of Mines – special dispensation was required because of his age, but this was obtained by the professor of mathematics who was struck by the youth's abilities.

On graduating two years later, Fourneyron undertook a series of commissions that were intended to broaden his education and equip him for a career in industry, and in 1821 he was given responsibility for investigating the use of coal for smelting iron ore at the iron works at Pont-sur-l'Ognon. (French industry was still labouring with traditional smelting methods involving the use of charcoal.) Here he began experimenting with cast and wrought iron, and in 1823 oversaw the construction of a rolling mill on the site. The mill was driven by a water wheel, and it was his involvement with this machine that marked the beginning of his lifelong interest in hydraulic motors.

Water power was of major importance in France at the beginning of the nineteenth century. A long series of wars had left the country in poor economic shape, and it was desperately struggling to keep up with Britain, which was forging ahead on all fronts.

Britain was a truly formidable rival. It led the world in industrial innovation, particularly in the important fields of heavy engineering, chemicals and textiles. In addition, it had vast reserves of coal and its industry was rapidly converting to steam. Early performance limitations on the steam engine had been largely overcome and costs were coming down.

France, by contrast, was an essentially agrarian society with a poorly developed economy. Much of its industry still relied on old and inefficient processes. What is more, France had desperately little coal. Its one saving feature was its water resources: the country was criss-crossed by rivers and waterways, and these afforded vast potential for water power.

The 'need' is clear. This combination of factors encouraged an intense interest in water power and it is this that accounts for the dominant role played by French academics in advancing hydraulic theory, and French engineers in the development of ever more efficient hydraulic machinery. An indication of the importance placed on water power was the substantial cash prize that was offered, in 1826, by the Société d'Encouragement pour l'Industrie Nationale to anyone who could produce a more efficient hydraulic motor suitable for large-scale application. Another indication was the growing use of the term 'houille blanche' – literally 'white coal' – to describe this vital natural resource.

Fourneyron began his quest for a better hydraulic motor in the early 1820s after carrying out a thorough study of the theoretical works of Bernoulli, Euler, Borda, Poncelet and others. After four years of experimentation he constructed his first water turbine in 1827. It was based on an original idea by one of his former teachers at the School of Mines in Saint-Étienne, Claude Burdin.

Burdin was a highly creative individual, but something of a failure as an innovator. He devised many novel hydraulic motors, some of which were built and tested. But his machines were complex (see, for example, Figure 2.13 overleaf) and gave only mediocre performance. They were never produced commercially.

His two lasting contributions to the field of water power were the concept of a horizontal outward-flow wheel – the type that Fourneyron perfected – and the term 'turbine', which he first used in a paper submitted to the Paris Académie des Sciences in 1822. The idea for the wheel probably came from a primitive form of horizontal wheel called a 'tub' wheel which evolved in Europe around the sixteenth century, and from the reaction wheels that Euler had worked on. The term 'turbine' was derived from the Latin 'turbo', which means 'a spinning top'.

Fourneyron's experimental turbine was installed in the bottom of an open flume at the Pont-sur-l'Ognon iron works and developed 4.5 kW on a head of just 1.4 metres (see Figure 2.14).

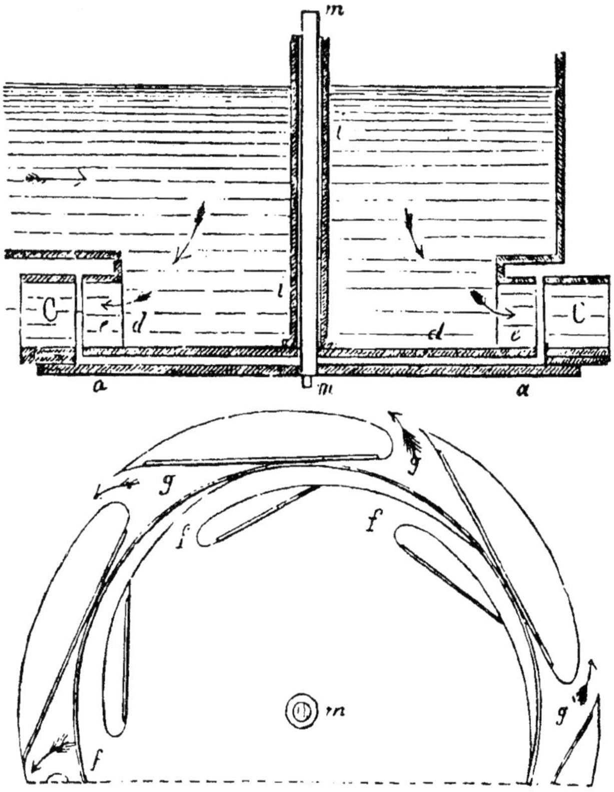

Figure 2.13 Burdin's design for a six-bladed outward-flow radial turbine

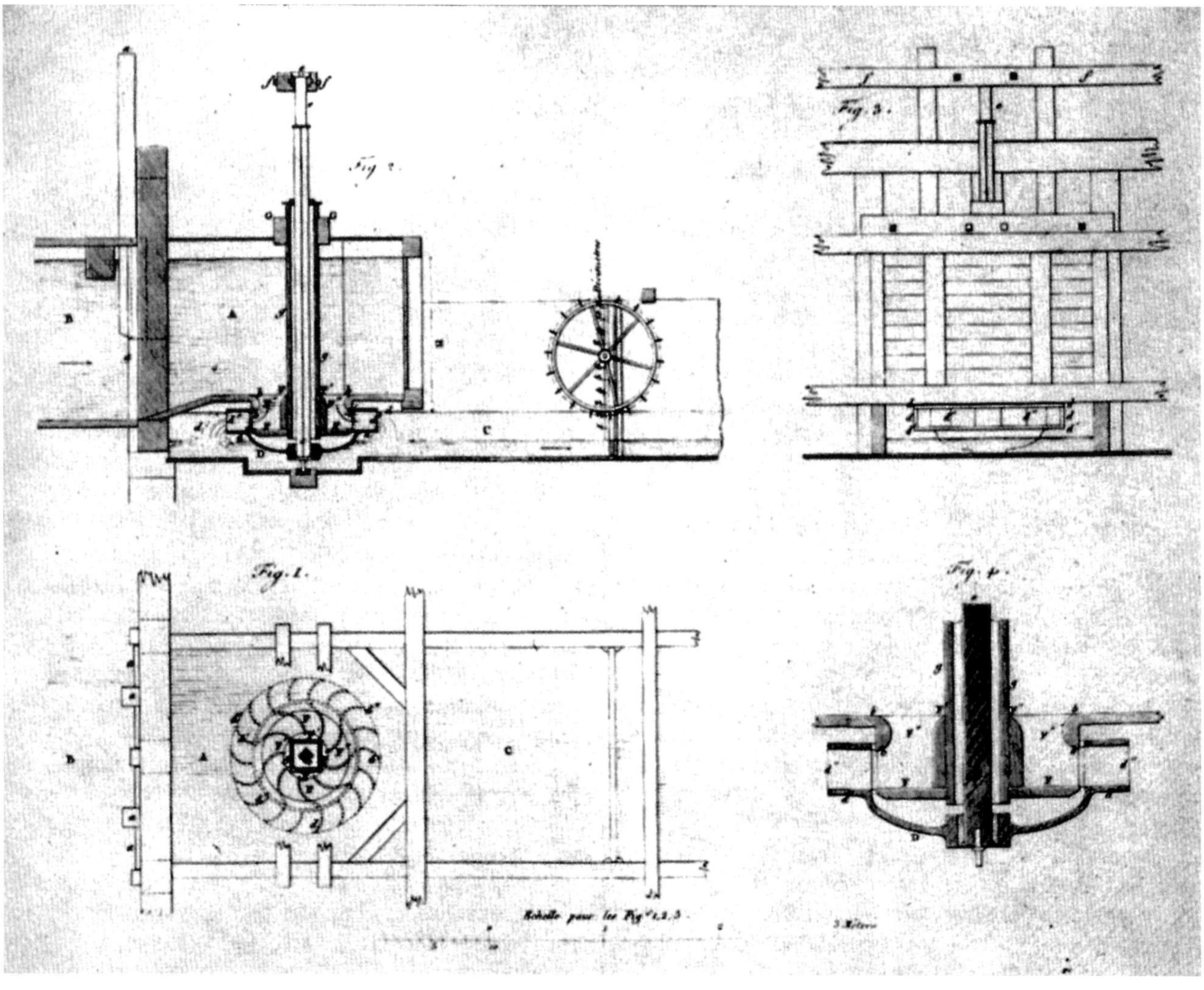

Figure 2.14 Fourneyron's experimental turbine at Pont-sur-l'Ognon

At the heart of the machine was a small horizontal runner with curved metal blades which was connected directly to the main drive shaft. The wheel was very similar in form to that suggested by Poncelet in a lecture in 1826. The only significant difference was that the flow pattern was reversed, with the water flowing from the centre of the wheel outwards, rather than the other way round. Water was directed onto the blades of the runner by a set of guide vanes, and after imparting its energy to the revolving assembly it fell directly into the tail race below.

What so excited the imagination of fellow engineers and industrialists was the fact that Fourneyron's turbine not only turned at speed – some 60 revolutions per minute – but that it also converted over 80% of the available energy into useful mechanical work. What is more, it could operate submerged and was not therefore significantly affected by flooding.

But Fourneyron did not stop there. He immediately began work adapting and modifying the machine to the requirements of the large iron foundries and textile mills. He was guided in this work by both theory and experiment. He made extensive use of models and even employed his own model-maker. The diligent approach that he adopted, and the breadth of the knowledge that he applied, were summed up by one of his biographers in the following terms:

> *He never installed a machine without first being assured of its satisfactory performance which he proved by repeated tests. As each successive machine was produced, in it were incorporated the fruits of his recent studies and experiments. Also, it usually embodied some novel means of overcoming an earlier difficulty. The number of experiments he made was enormous and embraced a wide field of activity, including strengths and properties of materials, practically all branches of hydraulics and fluid flow not to mention his earlier metallurgical work.*
>
> (Keator, 1939)

Fourneyron obtained his first patent in 1832, and in the following year he prepared a paper outlining a complete theory for his new device, in which he drew together threads from the work of several of his distinguished predecessors. It was this paper, together with an account of his turbines and their performance, that won him the prize offered by the Société d'Encouragement pour l'Industrie Nationale.

Fourneyron was also responsible for introducing many novel features in his machines. He developed cast-iron casings and high-speed gearing and introduced control gates which could be raised and lowered by hand to control the flow to the runner and thus the output from the machine. He also pioneered the use of high heads.

In 1837, he installed two turbines at a spinning mill at Saint-Blaise which operated on heads of over 100 metres, far in excess of anything previously attempted. The machines turned at the very considerable speed of 2300 revolutions per minute, each developing 45 kW with a maximum efficiency of around 80%. The runners on these machines were only 32 cm across and had a mass of less than 18 kg – a considerable contrast with the huge industrial water wheels that were then in vogue.

In Section 2.4 it was noted (in the extract from Hunter's notes) that wooden overshot wheels had a mass of about 700 kg kW^{-1}.

Before beginning the venture, he had had to build and test sections of wrought-iron pipe to prove to sceptical colleagues that the unprecedented high pressures of over ten atmospheres could safely be accommodated. In Section 5, I'll explain where this pressure comes from.

His largest machines were built at a textile mill in Augsburg. They operated on a head of 5 metres and developed 165 kW. Power from two of them was used to drive 30 000 spindles and 800 looms.

Fourneyron's machines embodied many of the features of the modern reaction turbine. They were:

- compact, which meant that they required less material to construct than a water wheel of equivalent output and were easier to install;
- highly efficient, which economized on water use;
- fast turning, which reduced or eliminated the need for gearing;
- able to operate submerged, which meant higher availability.

Furthermore, in principle they could be:

- designed to operate with a considerable head of water;
- built with much greater capacity.

Fourneyron continued experimenting with and improving his turbines right up to his death in 1867, by which time more than 100 machines had been brought into operation.

It is a testament to his success as an inventor and innovator that his ideas were so widely pirated and copied – despite the protection afforded by patents. He fought, and won, many legal battles in the courts over the infringement of his patent rights.

Fourneyron's turbines heralded the start of a new era of water power development which spelt the beginning of the end for the large industrial water wheel.

3 THE EVOLUTION OF THE FRANCIS TURBINE

3.1 James Francis at Lowell

As Chief Engineer and Superintendent of Locks and Canals at Lowell, James Francis had awesome responsibilities. It was his duty to ensure that the power needs of the main textile corporations were met – and Lowell was one of the USA's main textile centres in the mid-nineteenth century. At its zenith it boasted more than twenty large textile mills.

As shown in Figure 2.15, Lowell was strategically placed on a bend in the Merrimack River by the Pawtucket Falls and was close to Boston. It was served by a large canal, which had been built to enable ships to negotiate the rapids, and was supplied with sufficient volume flow to enable water wheels to operate unrestricted throughout much of the year.

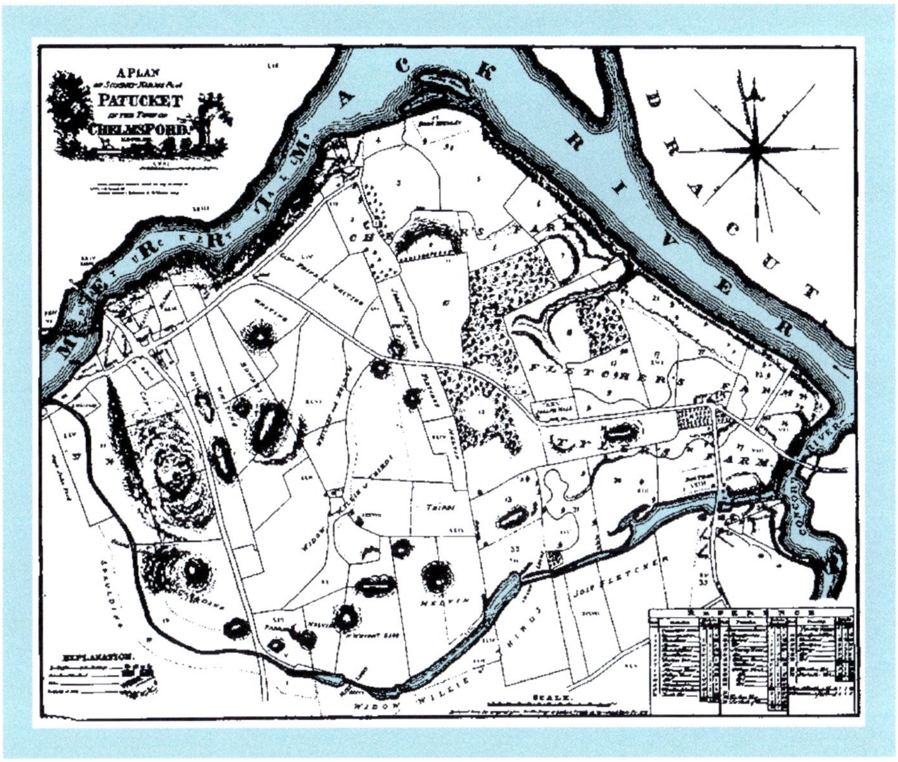

Figure 2.15 Map of Lowell, before industrial development

By the 1840s, when Francis came on the scene, the available head at Lowell had been increased to more than 10 metres by damming the river above the falls, and a succession of branch canals had been dug to enable greater use to be made of the power available. Francis continued this work. He enlarged and improved the network of canals so as to bring the supply system close to the practical limits of its capacity. He also organized the construction of more dams and reservoirs to control flooding and to provide a more constant and secure supply of water.

This in itself was no mean feat, given the size and complexity of Lowell's water system and the almost complete absence of precedent for many of the difficult technical problems that had to be overcome.

But Francis did much more: he pioneered techniques for controlling and measuring large volume flows of water and developed accurate methods for testing full-scale hydraulic motors. He was responsible for advising the textile corporations on the construction and operation of their machines and on many other matters relating to the use of water power in their factories.

Francis also conducted careful and exhaustive experimentation on several commercial water turbines, suggesting ways in which their performance might be improved. He carried out fundamental studies into turbine design, investigating, for example, the paths of water particles through different configurations of guide vanes and turbine runners. As a result of extensive studies, which were carried out over many years, Francis formulated empirical rules on the proper proportioning of hydraulic turbines. He later described this extremely detailed work in a major report, *Lowell Hydraulic Experiments*, which was soon acclaimed as a classic treatise on hydraulic engineering. The book, first published in 1855, went through several editions.

3.2 The work of Uriah Boyden

The first water turbines built at Lowell were outward flow turbines of the Fourneyron type. They were designed not by Francis, but by one of his great contemporaries, Uriah A. Boyden, a consulting engineer from Boston. But even his designs were derived from a turbine built by another distinguished US engineer, Ellwood Morris, who in turn had based his machine on rough sketches and descriptions of Fourneyron's turbines. He had had to rely on these indirect means because Fourneyron was a shrewd businessman who was careful not to divulge details which could be of value to potential competitors.

Morris subsequently commissioned local machine shops to build more turbines and these were sold to industrial firms in and around his home base of Philadelphia.

Morris's example was followed by others and by the mid-1840s a handful of outward-flow turbines were installed and running in the eastern USA. But the success of the Fourneyron turbine in the USA was primarily due to Uriah Boyden who, guided by sound engineering principles and mathematical analysis, refined and improved the design almost beyond recognition. Boyden's machines eventually outperformed those from Fourneyron's own workshops.

Boyden's turbines were machined to close tolerances and were true reaction turbines, that is, the wheels turned as a consequence of both the pressure of water on their blades and the impulse generated by its change of direction. Fourneyron had not been able to work to such high levels of precision.

Boyden took great care to determine the form and proportions of his wheels, producing full-sized technical drawings of the major components and conducting elaborate experiments on the test rigs to confirm his detailed calculations. Every effort was made to secure mechanical and hydraulic perfection. He also used the best materials available and insisted on the highest standards of machining and assembly – factors which combined to produce high performance and long lifetimes.

But Boyden did more than simply optimize the design; he also introduced a number of significant innovations that became standard features in many later turbines. These included a cone-shaped flume with fixed guides, which gradually increased the speed of the water and imparted a swirling spiral motion to it prior to its entering the runner; an improved gate mechanism for regulating the flow of water; and a diffuser which reduced the exit speed of the water and squeezed up to 3% more energy out of the flow (see **▼Diffusers and draft tubes▲**). By suspending the whole runner assembly from overhead, he was also able to avoid the problems with footstep bearings (thrust bearings used to support the lower end of a vertical shaft) that had so dogged the performance of many of Fourneyron's early machines. (The idea of suspending the runner was not new, but Boyden was the first to produce a successful bearing arrangement.) Some of these features can be seen in Figure 2.16.

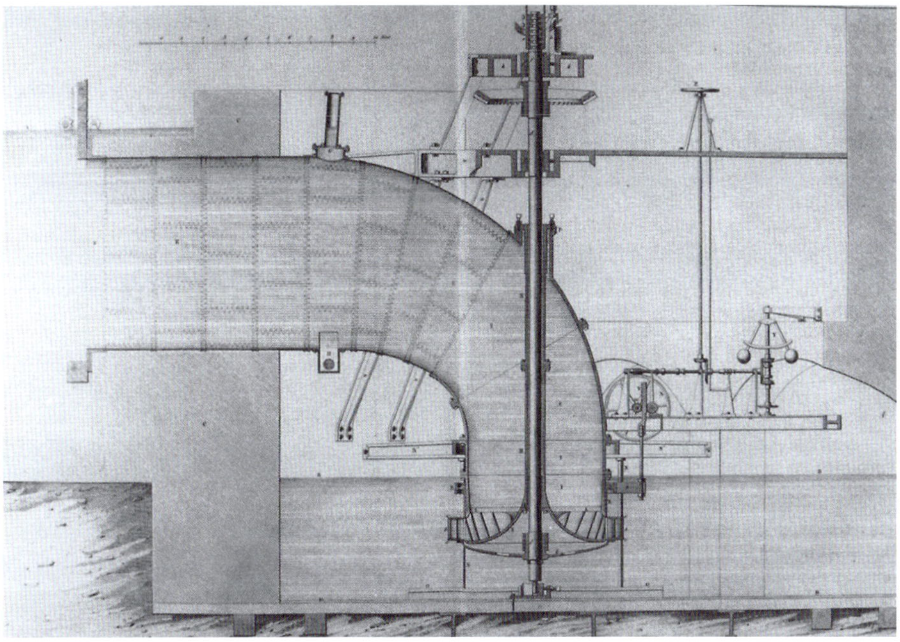

Figure 2.16 Tremont (Boyden–Fourneyron) turbine in cross-section

The first Boyden–Fourneyron turbine was installed at the Appleton Mills at Lowell in 1845 and developed a modest 55 kW with an efficiency of 78%. Later machines reached unprecedented sizes – some had wheels almost 3 metres across and developed over 500 kW, three times the output of Fourneyron's largest machines. The fact that turbines were not affected by backwater problems meant that they could be set lower and utilize a greater head than large breast wheels. This gain, together with superior efficiency, gave a total improvement of energy extraction of up to 40%.

Boyden's more scientific approach and his care and attention to detail soon established him as the USA's foremost water turbine designer. Sale of his patents earned Boyden a fortune which enabled him to retire early and pursue a lifelong interest in pure scientific research.

▼Diffusers and draft tubes▲

Boyden's diffuser consisted essentially of two stationary discs placed concentrically around the wheel, the internal diameter of the discs being slightly larger than that of the wheel, and the external diameter about twice that of the wheel. The vertical distance between the discs increased with diameter, as shown in Figure 2.17.

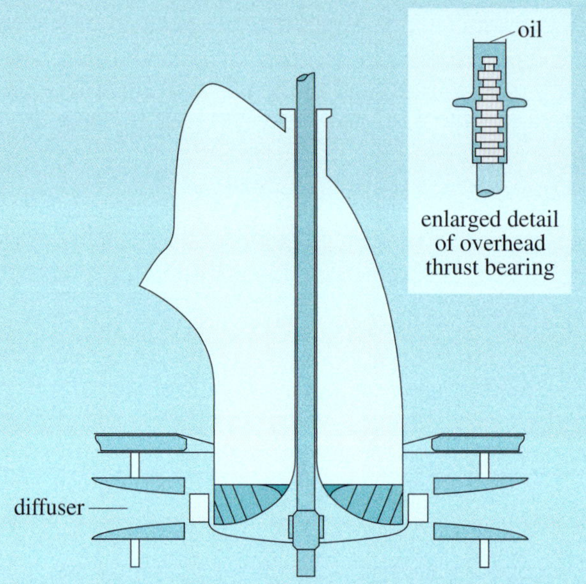

Figure 2.17 Boyden wheel and diffuser, in cross-section

The diffuser rendered useful a part of the power that was otherwise lost as a consequence of the water leaving the wheel with a considerable speed. The effect of using such a device was the same as if the head of water acting on the turbine were to be increased – a partial vacuum was created at the output of the runner because the increasing cross-section in the draft tube forces the water to decelerate as it moves away from the runner, but the inertia of the decelerating water 'pulls' on the water behind it.

Diffusers were used in most Boyden turbines from the mid-1840s onward, although they were dispensed with in later machines because of their great size, which meant that very large and expensive wheel pits had to be constructed to accommodate them. They were, nevertheless, an extremely important invention which anticipated the flaring draft tube of the modern reaction turbine.

Draft tubes actually pre-date the diffuser. They were made from wood or sheet metal and were little more than airtight tubes. The idea was conceived by an American, Austin Parker, in 1833, and was subsequently developed by his brother Zebulon and patented in 1840.

Draft tubes (effectively acting in the same way as a siphon) enabled the turbine to be raised above the tail race without loss of head. This permitted easier access for maintenance and repair, and later facilitated the switch from vertical-shaft turbines to horizontal ones that could be coupled directly to machinery, eliminating heavy and expensive bevel gearing.

Flared draft tubes (see Figure 2.18), which combined the function of the early draft tube and the Boyden diffuser, came later. However, they were more expensive to engineer than straight-sided versions, and the extra 2–3% efficiency improvements were often not considered sufficient to justify the extra expense. They did not really come into their own until the development of hydroelectricity towards the end of the century and the introduction of reinforced concrete, which enabled complex shapes to be engineered relatively easily and cheaply. Today, flared draft tubes are almost universally used in reaction turbines.

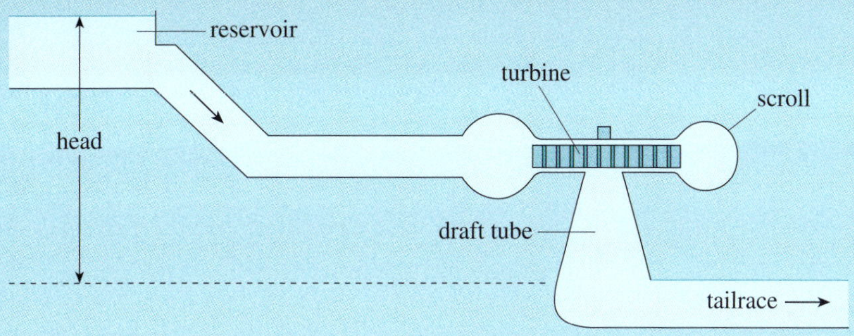

Figure 2.18 Flared draft tube

3.3 The original Francis turbine

The Locks and Canal Company purchased the rights to manufacture Boyden turbines in 1848, and Francis became involved in adapting these to the specific needs of the large textile mills at Lowell. With the help of experimental data (largely provided by Boyden, who for a time collaborated on the work), Francis formulated general rules on the correct sizing of the various turbine components for given conditions of head and flow, and successfully applied these rules of thumb to each new machine that was installed.

His approach, like that of Boyden before him, was to trace the path of water particles through the passageways of the turbine, laboriously calculating the area of the channel and the speed of flow every half inch or so. The results were then plotted out on paper at full scale (see Figure 2.19). Francis was careful to see that any bending or twisting of the water stream was kept to the barest minimum.

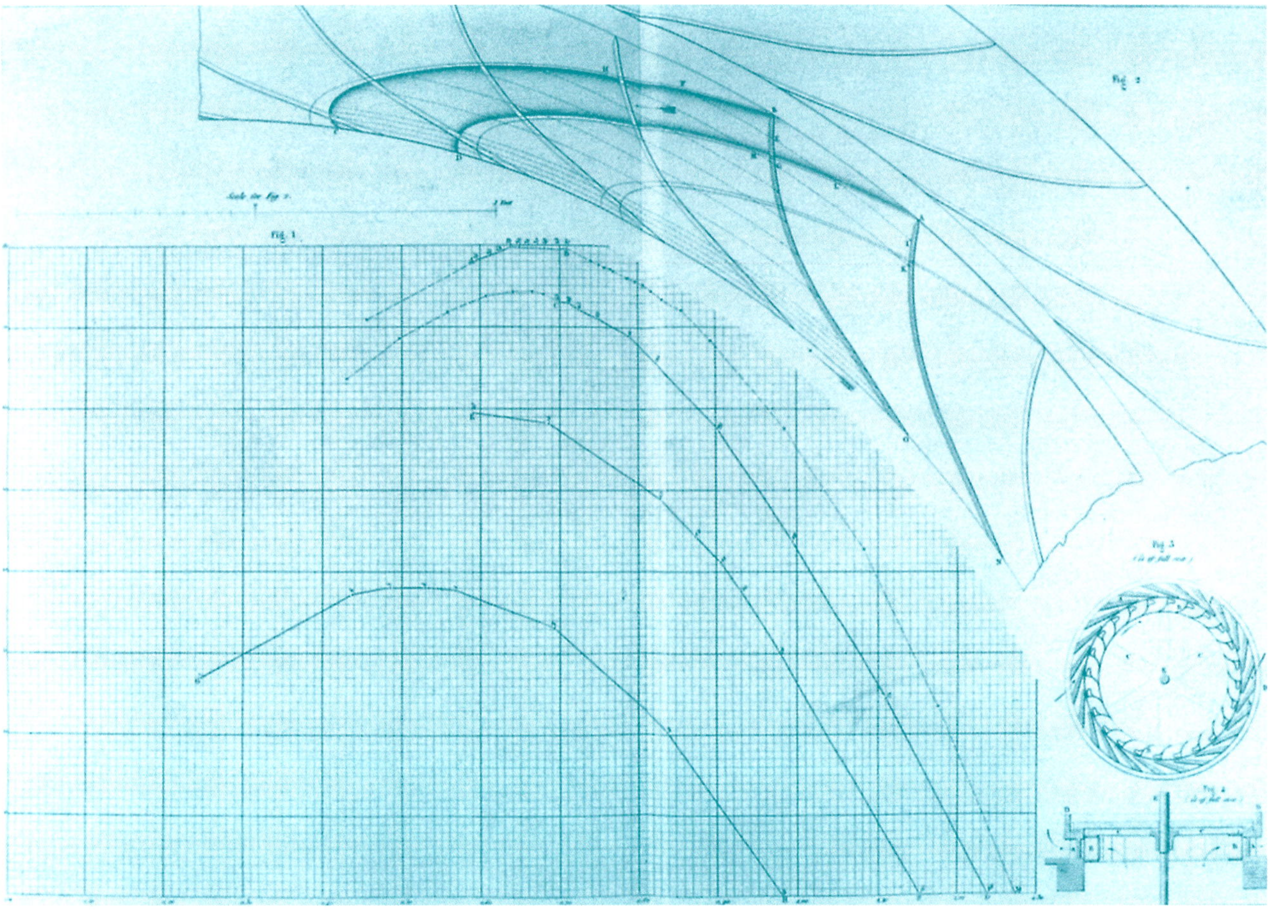

Figure 2.19 Absolute path of water flowing through a Francis turbine. In the inward-flow design, the channels provided by adjacent buckets (shown in cross-section near the top right of the figure) narrow as the water approaches the inner rim. Unless compensated for by increasing the depth, this would accelerate the water. The absolute path is shown graphically

Francis also developed his own rigorous tests and procedures and accurate techniques for measuring turbine efficiency. It was on this unique blend of scientific style and meticulous experimental work that Francis's reputation was built.

Francis became the first individual to carry out scientific tests on full-sized turbines when in 1851 he investigated the performance of a Boyden wheel built for one of the Lowell mills. For this he made use of a simple friction brake that had been invented in the 1820s by Prony, a French engineer.

The importance of ▼**Prony's dynamometer**▲ was that it enabled the performance of wheels to be measured to an accuracy of 2–3%, so that differences in design could be assessed quantitatively.

Large-scale testing had become necessary at Lowell since, under the terms of the contract with Boyden, the Locks and Canal Company was obliged to pay compensation on a sliding scale depending on performance. Increments were due for each extra percentage point improvement over and above an agreed baseline figure. Accurate techniques for measuring large-volume flow were also required so that the power company could ensure that mill owners were complying with regulations (and calculate excess charges if they were not).

▼Prony's dynamometer▲

Prony's dynamometer (or friction brake, Figure 2.20) measured the useful work performed by the prime mover by causing the whole of that work to be expended in overcoming the friction of a brake (hence the term 'brake horsepower'). It had shaped wooden brake blocks which were clamped around the drive shaft of the wheel and held tightly in position by a metal strap. The brake assembly was attached to a beam, which was maintained in a horizontal position against the wheel's torque by weights placed in a scale pan. The mechanical effort was calculated from the rotational speed of the shaft, the weight supported in the pan, and the horizontal distance of the weight from the shaft.

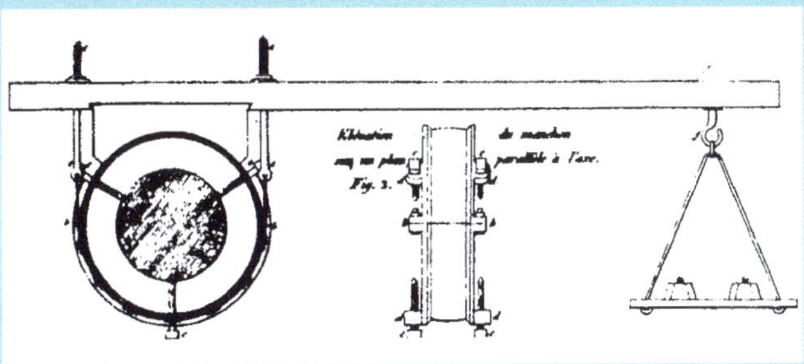

Figure 2.20 Prony's dynamometer

Thus Francis and Boyden were amongst the first to introduce testing as a part of the proving process for new machines before sale to the customer. This important development had actually been pioneered by Ellwood Morris, who had first imported the Prony brake into the USA in the 1840s. With an accurate dynamometer Morris was able to demonstrate the performance of his turbines to potential customers and to predict the performance they could expect under specified conditions of head and flow. In mid-nineteenth-century America, this was a wholly novel development.

Demand for outward-flow turbines grew apace, keeping Lowell's engineering workshops busy. By 1858, Francis had supervised the design and installation of over fifty machines with a combined operating capacity of more than 900 kW.

But there were problems with the Boyden–Fourneyron turbine of which Francis had been aware from the beginning. Paramount amongst these were its low efficiency at reduced flow rates and its susceptibility to damage from floating debris. Power output dropped off dramatically when the flow rate fell below the design capacity – at half the design capacity efficiencies as low as one-third were common. Furthermore, performance deteriorated over time as flotsam, drawn into the turbine, bent and distorted the finely engineered guides and runner. Average efficiencies fell by as much as 10% in the course of a couple of years.

Francis also recognized that there was considerable scope for both increasing the capacity of water turbines relative to their size, and increasing their shaft speeds for a given head so as to reduce the need for gearing. He was aware, too, of the need to reduce costs. Boyden had spared no effort to achieve mechanical perfection and his turbines were, if anything, over-engineered.

He specified high standards of machining and assembly, and used only the best materials. For example, he often used bronze for the guide vanes and runner blades.

Amongst the various turbine types that were beginning to appear on the US market in the late 1840s, Boyden's machines could be compared with the custom-built Rolls-Royce of today. A less elaborate, more robust machine was needed, one that could be produced in numbers and sold at a more competitive price.

Francis was not the man to produce such a machine – he was more of an improver than an inventor. However, he did point the way to the eventual fulfilment of this need. After careful evaluation of a number of commercial turbines – including those of Fourneyron and Boyden – he turned his attention to a lesser-known turbine that had been designed by a New York millwright, Samuel Howd (see Figure 2.21).

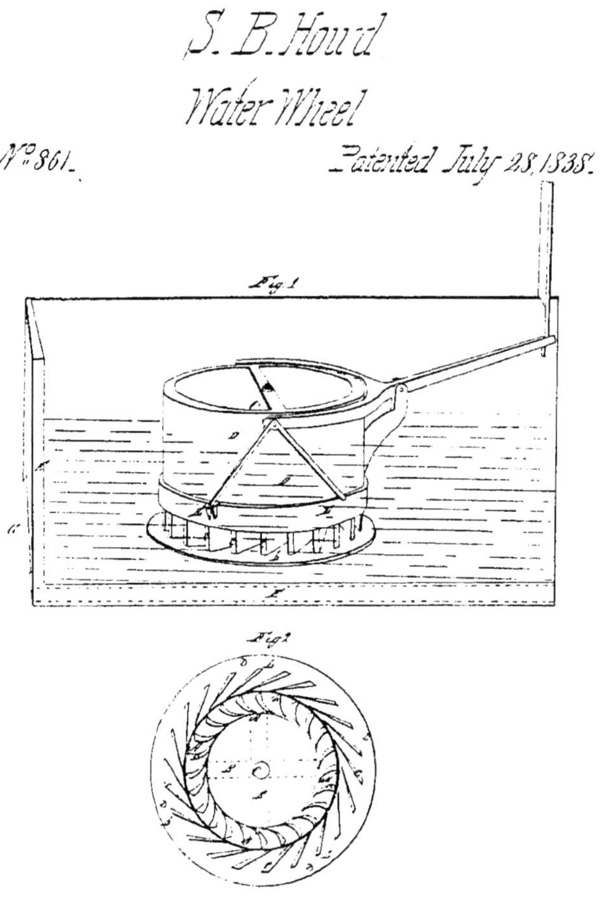

Figure 2.21 Drawings for the Howd patent, 1838

Francis concluded, after much theoretical analysis, that Howd's 'centre-vent' design offered more genuine scope for improvement than did the outward-flow configuration of the Boyden–Fourneyron turbine. Francis wrote that he was 'led to this design by the consideration of the path of water passing through the wheel'. Edwin Layton has suggested that:

> *What Francis probably meant was a matter of rather simple geometry that his and Boyden's methods of design would have made very obvious. The outer circumference of a turbine wheel is longer than its inner circumference. Thus in an outward-flow turbine the water would tend to spread and eddy as its channel*

> *widens, a very undesirable result for anyone seeking to lessen water resistance. Reversing the direction of flow offered an obvious remedy.*
>
> (Layton, 1979)

Thus it was Howd's original machine, patented in 1838, that became the blueprint for the Francis turbine (see Figure 2.22). The wheel turned on a vertical shaft like that of the Boyden turbine. It was made from cast iron and mounted in a wooden flume. Guide vanes, set around its circumference, directed the flow inward over the wheel to a central discharge. This arrangement generated a vortex so that the speed of the water flowing onto the wheel was closely matched to the rotational speed of the blades. This was the essence of Howd's invention.

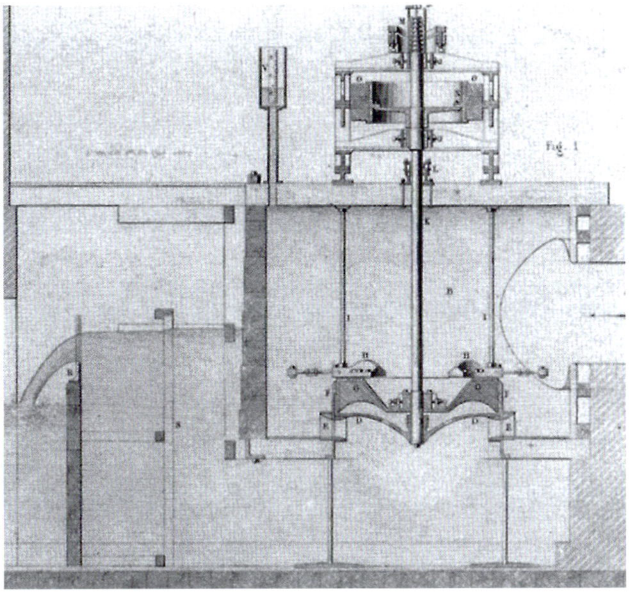

Figure 2.22 The Howd–Francis centre-vent turbine

There was very little that was actually new in Howd's device. He probably based the original design on a primitive form of horizontal impulse wheel called the **▼Flutter wheel▲**, which was modified so that water struck it from all sides. This arrangement was similar to that proposed more than a decade earlier by Poncelet as a means of improving the efficiency of his rationally designed undershot wheel. Guide vanes were also known – Fourneyron had used them in his turbines in the late 1820s, although Howd was probably unfamiliar with this work since details had not been published in the USA at the time.

The essential ingredient contributed by Howd – his 'inventive step' – was the concept of using radially disposed guide vanes to produce a vortex around the spinning runner so that the inflowing water could enter without shock – one of the necessary conditions for an efficient hydraulic motor. The machine itself was actually quite crude in construction, certainly by comparison with the sophisticated Boyden wheels; there was no governor – output was controlled manually by raising and lowering a cylindrical gate using a lever.

The Proprietors of Locks and Canals at Lowell obtained a licence to manufacture and sell Howd machines, and Francis spent many months experimenting and adapting the design. He built and tested his first model turbine in the Lowell workshops in 1847, recording an efficiency of 69%. The runner for this machine was similar in construction to that of Boyden's early wheels, but was designed for inward flow. Two years later, after further tests, he built two full-sized wheels at the Boott cotton mills at Lowell.

Each machine developed 185 kW with a conversion efficiency of 79%.

▼Flutter wheel▲

The Flutter wheel was a simple horizontal water wheel that turned under the impulse from a stream of water in much the same way as the undershot wheel. It was a direct copy of the early Norse mill which appeared in Northern Europe sometime during the Dark Ages (see Figure 2.23). The water fell from a chute onto flat paddles set at an angle around a thick wooden axle. Immediately above the wheel, and connected to it by a rigid shaft, was the millstone which turned with the wheel. This whole assembly rested on a simple bearing set into the top of a pivoted wooden beam. The position of the beam could be adjusted by means of a hand screw or a simple wedging device, so that the millstone could be raised or lowered to regulate the fineness of the flour.

Flutter wheels were popular in the American backwoods in the early part of the nineteenth century. They did not require gearing, and were therefore easy and cheap to build. However, they were also highly inefficient. The best achieved efficiencies of only about 20%.

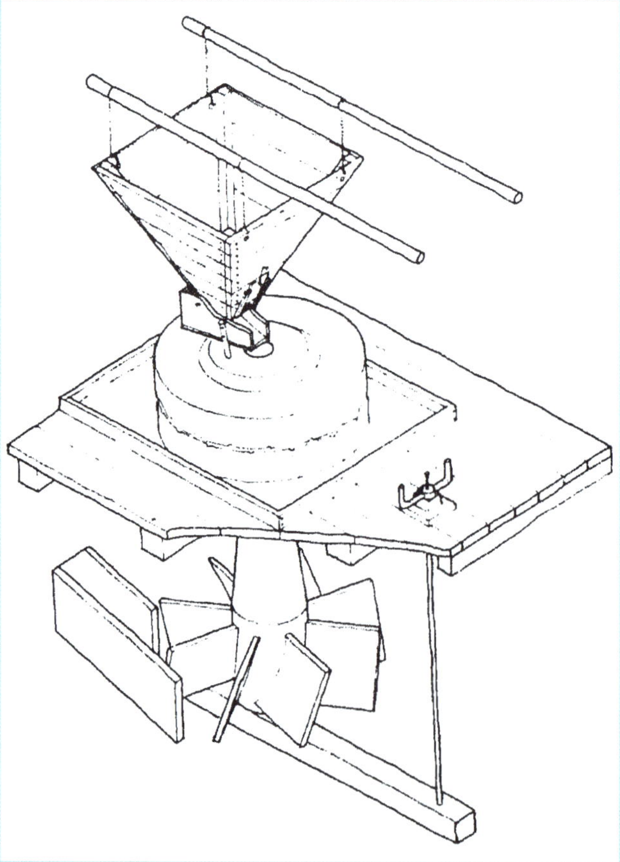

Figure 2.23 A Shetland Isle version of a Norse mill

The final form of the Boott wheels bore little resemblance to Howd's original. This is hardly surprising given the resources at Francis's command and the experience and data that Boyden had bequeathed him. Skilled mechanics and well-equipped workshops were also on hand.

Much to Francis's chagrin, his turbines were little used in North America, although they did find some application in Europe in the latter part of the century. It was not that his machines did not work. They did. However, they showed no performance advantage over the better-established outward flow turbines of Boyden or over some of the other types of water turbine that had by this time appeared on the market. They were also expensive to manufacture.

It is worth noting that Francis was not the only individual to investigate inward-flow turbines in the 1840s. James Thomson, Professor of Engineering at Belfast University and brother of Lord Kelvin, patented a 'Vortex' turbine in 1850. Later versions of this machine had scroll cases, pivoted guides and centre-discharge wheels, and were probably the first machines to incorporate all the essential features of the modern Francis turbine. Thomson turbines were manufactured in the UK for nearly 80 years.

Despite this commercial failure, Francis's achievements in the field of hydraulic engineering were considerable. He made his mark on a whole generation of engineers. This was no mean feat given the state of engineering in mid-nineteenth-century America. The fact that the population was small and thinly spread placed expertise at a premium. It numbered only 10 million people in 1850, half the number in Britain, and a large proportion comprised unskilled labourers who worked on the land in farming or mineral extraction. Furthermore, there was no established tradition of fine workmanship as there was in Europe. Engineering was rough and ready. It evolved to meet the needs of the frontier.

Hydraulic engineering, in particular, did not have the same professional status as it did in, say, France, where water power was of relatively greater economic importance and where the profession was well established. It was left largely to local carpenters, millwrights and foundry men. Promising young engineers were encouraged to go into steam or heavy engineering, and as a result competent hydraulic engineers were few and far between. In this context, Francis's achievements were all the greater.

But Francis's turbine did have obvious structural and operational advantages over the outward-flow turbines of Boyden and Fourneyron. The runner could be mounted inside a scroll casing, and guide vanes used to impart a swirling motion to the water as it entered the runner. The speed and direction of the water could be adjusted to match that of the revolving blade tips. The flow was less turbulent. Furthermore, a draft tube could more easily be attached to enhance the power output.

It was not surprising that other engineers soon began exploring the potential for inward flow.

3.4 The period of cut-and-try

Although Francis had reversed the flow through the wheel and thus set the direction (both literally and metaphorically!) for future development, he was unable to overcome some of the more fundamental problems associated with the Boyden turbine. His machines, like those of Boyden, had large-diameter wheels which turned relatively slowly; both types of machine were disappointingly inefficient at less than full flow and were prone to obstruction and damage from floating debris. Furthermore, neither Francis nor Boyden had been able to develop a truly satisfactory gate mechanism to control the flow onto the wheel. Their turbines were commonly equipped with plain 'rabbit trap' gates which were cranked up and down.

Hydraulic theory, which had been advanced over the years by so many distinguished individuals, and which had provided important insights into improving turbine design, offered no solution. The classical theorists like Bernoulli and Euler had, of necessity, made simplifying assumptions about the processes taking place within the hydraulic systems they studied. For example, they ignored the effects of friction and the internal resistance (viscosity) of water. The result was frequently that theoretical predictions were at odds with experimental observation. The problems were especially acute with turbines because of the complexity of the flow patterns associated with them.

Consequently, US turbine builders turned to intuition to find a way out of their difficulties. There followed one of the most extraordinary periods in the history of the water turbine. Empiricism returned with a vengeance, and held sway for a generation.

> *All the scientific teachings of Boyden and Francis were thrown to the winds, and the great god 'Cut-and-Try' came into his own. If a wheel did not come up to expectations, its buckets were chipped back, up, or down, or its blades pounded, until it gave something better.*
>
> (Safford and Hamilton, 1922)

Whereas Francis and his colleagues at Lowell had laboriously computed the mean path of water particles through the turbine's runner as a guide to formulating the best angles and curvatures, the new generation of turbine builders advanced their designs empirically, either by making small modifications to existing wheels or building and experimenting with scale models. They had available to them far more advanced metalworking equipment than earlier millwrights, as well as a broader background of production experience and greater financial resources.

> *Instead of making up a wheel by casting parts from the old, well-worn patterns drawn from the stockroom each time a turbine order came in, the new wheel builders were continually introducing this change or that in the shape, dimensions, or proportions, seeking improved performance, greater efficiency, or simpler or better methods or materials of production. Within a short span of years a new and distinctive class of wheels gradually emerged, each made with its own characteristics and peculiarities but reflecting the often common results of a widely shared experience, each builder influencing and, outside the uncertain limits of patent infringement, copying others. What eventually became known as the 'American' or 'mixed-flow' turbine gradually took form and rose to even higher levels of achievement.*
>
> *This class of turbine was distinguished typically by novelty and unconventionality in design; simplicity and durability in mechanical arrangements and equipment; quantity or small lot production in standardized sizes to stock rather than to order;*

avoidance of refinements in detail and materials; and, at every stage, low cost. Within a relatively few years the leading makes of turbine were performing with an efficiency which closely approached and then excelled that attained by the individually designed, precision-built, and costly wheels of Boyden and Francis.

(Hunter, 1979)

To increase the speed and output of the basic centre-vent wheel, it was necessary to reduce its diameter and to extend the blades downwards and inwards. This increased the area of blade that was subject to the pressure and reaction of the water and it enabled larger volume flows to be accommodated.

The effect of these changes can be seen in Figure 2.24 and in Table 2.1, which gives performance data for four equivalent-diameter (0.3 m) runners under a head of 9.15 m.

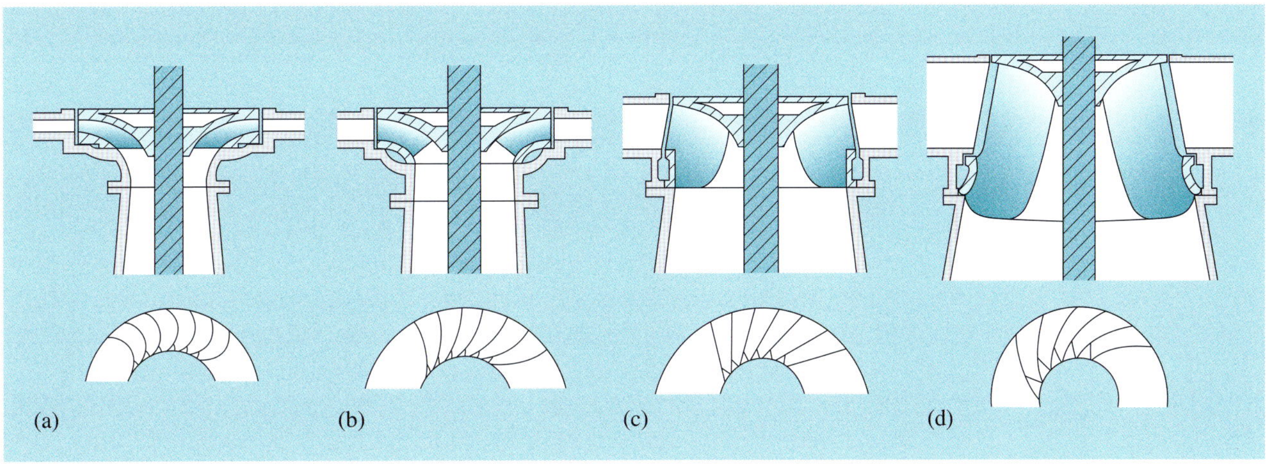

Figure 2.24 Development of the runner of a centre-vent wheel for increased power

Table 2.1 Measured performance data for four 0.3 m diameter runners under a head of 9.15 m

Runner type	Flow rate† / kg s^{-1}	Power output / kW	Revolutions per minute	Rotational speed / rad s^{-1}	Torque / N m	Efficiency / %
(a)	46.7	3.2	460	48.2	66.4	*
(b)	155.3	11.1	554	58.0	*	79.6
(c)	349.7	24.9	600	62.8	*	79.3
(d)	570.6	41.4	730	76.4	542.0	*

* See SAQ 2.1 below. † 1 kg s^{-1} ≡ 1 litre s^{-1}

SAQ 2.1 (Outcome 2.4)

Recall from Block 3 Part 2 the relationship between torque and power ($P = \Gamma\omega$) to complete the 'torque' column of the table. By thinking of the turbine as a converter of the gravitational potential energy of the water, complete the last column of the table.

3.5 The evolution of the mixed-flow turbine

The idea of combining radial inward flow with axial flow to produce a 'mixed-flow' turbine appears to have been suggested first by Francis, although he did not play any significant role in developing the concept. In fact it was Boyden who, in 1849, built and tested the first centre-vent wheel which incorporated a small element of axial flow.

The individual who did most to advance the mixed-flow concept was undoubtedly Asa Mathajer Swain, a skilled mechanic and foundry pattern maker, who for several years worked at Lowell on the development and testing of turbines. Swain built his first full-sized mixed-flow turbine at Lowell in 1859 after carrying out a series of tests on a six-inch model. He later went into business on his own, setting up the Swain Turbine and Manufacturing Company.

Swain introduced deep buckets into his wheel and took great care with the design of the water passages through the machine. Smooth surfaces and gentle curves reduced turbulence and frictional losses. The result was a turbine with excellent performance characteristics. Francis was impressed by the Swain wheel and encouraged its adoption at Lowell in the 1860s and beyond.

Swain did not work entirely without scientific input. Conscious of the need to proportion the water passages correctly to facilitate flow through the wheel, he engaged one of Francis's assistants, Hiram Mills. The suggestion that the height of the runner be increased appears to have come from Mills.

Swain later claimed that it was he who had actually invented the mixed-flow principle. But since he had not made this claim in his original patent of 1860 he was unable to press his case. The issue came to a head in 1874, when Swain's company took legal action to prevent another manufacturer and potential competitor from selling mixed-flow turbines of similar overall form. However, Swain's case rested on a claim that had been introduced into his original patent when it was reissued in 1872, some years after the concept of mixed flow had become widely adopted. Not surprisingly the US Supreme Court found against Swain, upholding the other company's case that a reissued patent could not be extended to cover claims not in the original.

Notwithstanding his legal failure, Swain's contribution to turbine development was considerable. His wheel anticipated some of the main features of the modern Francis turbine.

Thus Swain's turbine was the first of a new generation of more powerful, more robust and significantly cheaper prime movers (see Table 2.2). The more notable members of this generation included:

- the 'American' turbine, which was the first with standardized dimensions and interchangeable parts to be produced in quantity;
- the Risdon turbine, which attained an efficiency in excess of 90%;
- the amazing 'Hercules', which surprised everyone by developing three times as much power as other machines of equivalent runner diameter. The Hercules is also remembered because it turned at much greater speed than previous mixed-flow turbines and was the forerunner of the equally successful 'Victor' and 'Samson' machines.

Table 2.2 Characteristics of some American water turbines, 1849–97, on the basis of 30-inch (760 mm) wheel and 12-inch (300 mm) head

Type	Date	Maximum power output /kW	/horsepower	Efficiency (%) full power	half power
Francis	1849	0.15	0.20	79.7	55.0
Boyden	1851	0.35	0.47	79.4	46.0
Swain	1869	0.44	0.59	82.2	67.7
	1874	0.57	0.76	83.5	72.1
'American'	1872	0.31	0.41	76.8	72.0
Risdon	1873	0.38	0.51	90.5	73.0
'Hercules'	1876	1.07	1.44	83.8	73.0
	1897	1.33	1.79	84.0	70.5
'Victor'	1878	0.98	1.32	86.7	65.0
	1897	1.34	1.80	80.6	68.8
'Samson'	1897	1.38	1.85	82.0	75.6

The 'Hercules' turbine was designed by a Pennsylvanian mechanic, John B. McCormick, and based on a large-capacity wheel called the 'Little Giant' that had been designed five years earlier by two brothers, John and Matthew Obenchain. The buckets in these wheels actually protruded below the bottom rim so that the water discharged downwards and outwards into the draft tube. This same feature can be seen in the 'Samson' turbine shown in Figure 2.25 overleaf.

McCormick, who had no academic training, demonstrated an unusual flair for turbine design. He is said to have whittled out his wooden pattern of water passages until 'they looked right'. The story goes that he:

> *... desired to economise and lessen the discharge of an old inward flow turbine in a small mill that had come into his possession, and so started experiments with some spoon shaped extensions to the buckets of the old wheel, which gave such good results that he made patterns for a new turbine, cast his turbine in the iron foundry, [and] finished it off on the lathe ... Not being bound by precedent, he had given his runner much greater depth and relatively smaller diameter than that found in previous turbines, and he put more of the discharge through the outward flare of the bucket than previous experimenters had dared to do. It has been suggested that the remarkable success of his first model was more by happy accident than by way of superior knowledge.*
>
> (Freeman, 1924)

The combination of improved design, quantity production and fierce competition inevitably drove down the cost of reaction turbines. One engineer later summed up half a century of advance in the following terms: 'equal power in one-half the space, at one-fifth the cost, and with much simpler construction' (quoted by Hunter (1979)).

The new turbines represented a major advance over the large industrial water wheels they displaced. Table 2.3 shows just how much better they were in

Figure 2.25 Runner of a 'Samson' turbine

respect of their size, weight and power-to-weight ratio. It compares a high breast wheel with the 'Victor', one of the new generation of machines, operating under comparable conditions of head and flow.

Table 2.3 Comparison of the characteristics of a high breast wheel and a 'Victor' turbine

	High breast water wheel (1886)	'Victor' water turbine (1882)
Head / m	4.65	5.55
Flow rate* / $m^3\ s^{-1}$	1.0	0.8
Power / kW	39.4	36.6
Diameter / m	7.0	0.76
Width or height / m	2.5	0.7
Mass / kg	14 143	534
Power-to-weight ratio / W kg^{-1}	2.8	68.5

Source: adapted from Reynolds (1983)

* $1\ m^3\ s^{-1} \equiv 10^3\ kg\ s^{-1}$

What eventually brought this extraordinary period of turbine development to an end was the introduction of hydroelectricity in the closing years of the century. The new power companies required much larger machines that were not only capable of turning at very much higher speeds, but which also performed to precise technical specifications. Without a deeper theoretical understanding of the internal workings of their empirically advanced turbine designs, US manufacturers were unable to respond to this challenge. Cut-and-try methods were wholly inappropriate.

Before I go on to discuss the influence of hydroelectricity on the development of the American mixed-flow turbine – or the Francis turbine as we now call it – there is one other development that is worthy of note. This is the establishment of commercial testing facilities in the USA in the 1870s.

3.6 Commercial testing

Commercial testing led to the rapid dissemination of accurate and impartial information on turbine performance, and this had two important effects. For the first time, the outcome of modifications could be fully quantified and independently assessed.

Performance tests on full-scale machines did not, of course, displace or supersede model testing as a means of perfecting design. They were complementary. It was – and still is – much cheaper to experiment on physical or computer-based models before committing substantial investment to building full-sized machines for the market.

The man largely responsible for introducing commercial testing was James Emerson, an outspoken eccentric with an abrasive character and strongly held views. Emerson was originally engaged by Swain in 1869 to assist with wheel testing at Lowell. But three years later he moved to Holyoke, Massachusetts, to set up his own commercial testing flume (see Figure 2.26). He was appalled by the poor quality of some of the wheels on the market and challenged manufacturers to have their machines tested by him to prove their claims – a service for which he naturally charged a fee.

The results of Emerson's tests were published in his monthly magazine *The Turbine Reporter* and in annual reports which were available to anyone on request. This did much to educate the public and reduce confusion resulting from rival manufacturers' often exaggerated claims. Within less than a decade Emerson had determined the performance characteristics of all wheels of any repute whatsoever and made public his findings. He was one of the first consumer champions.

The venture proved a great success, so much so that it became common practice to base the contract of sale upon the performance of a wheel as determined in the test flume. Those manufacturers who refused Emerson's challenge found it increasingly difficult to sell their products. 'Many of the loudest mouthed', Emerson wrote in his fourth annual report, 'have been silenced and their wheels consigned to the "old Junk Shop" … the time has passed for running Turbine Water Wheels by wind'.

The Holyoke Water Power Company took over Emerson's test rigs some years later and developed his work, building a new, more advanced flume in 1881. Over the years, Holyoke established itself as a national test centre for water turbines, and acquired a reputation for rigour and accuracy. Eventually, the larger turbine manufacturers established their own facilities for

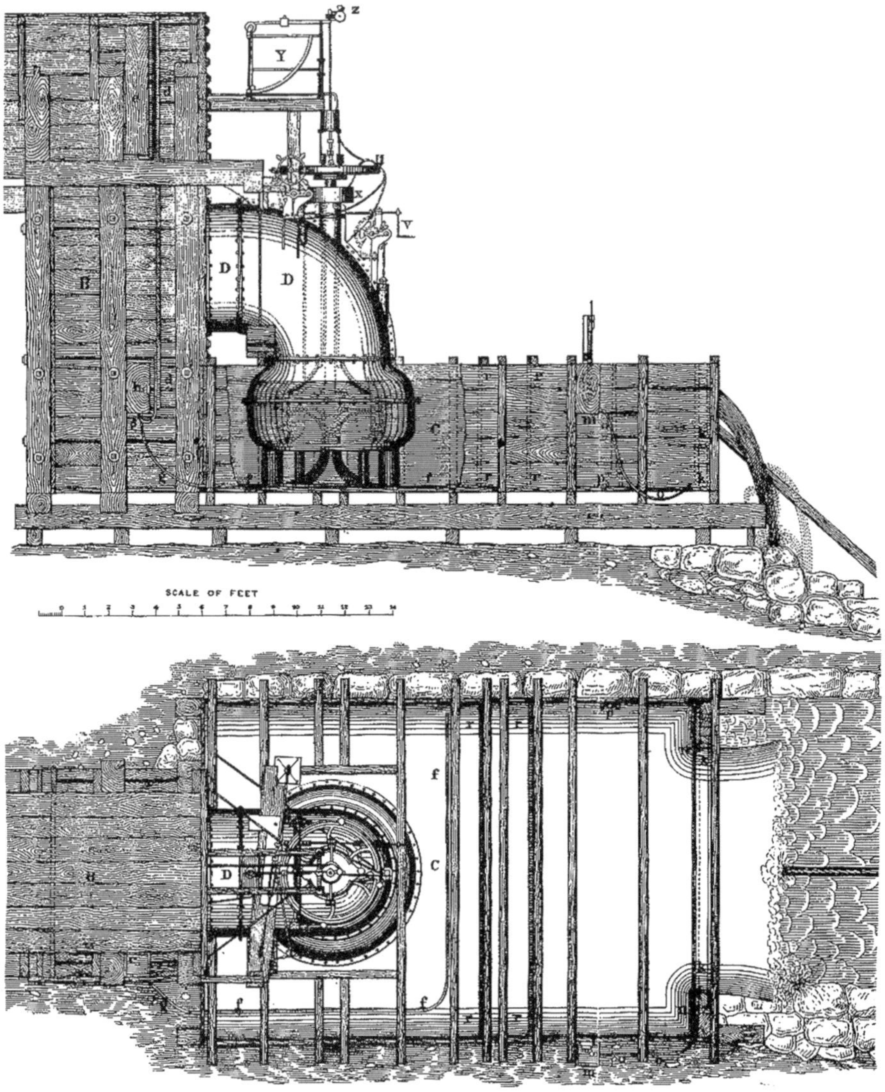

Figure 2.26 Emerson's testing flume at Lowell, 1872

monitoring the performance of their machines and evaluating the effect of design changes.

A number of engineers and historians of science have emphasized the significance of this development. The following quote by Chester Lamer, taken from an address to the American Society of Civil Engineers in October 1909, is typical.

> *In the evolution of turbine design in the USA, probably no factor has played a more important part than the Holyoke testing flume. Up to the present time, more than 1800 turbines have been tested there, and these tests have undoubtedly had a very important influence on the advancement of the art. The fact that all wheels (within certain limits of capacity) are tested there under essentially equivalent conditions, adds greatly to the value of the results, because it renders them directly comparable. Tests made in place are subject to the effects of so many local conditions that one scarcely knows when it is fair to make direct comparisons, but Holyoke tests are not subject to this uncertainty, because the conditions are known and constant.*

Much of Emerson's success was due to the use of a flume, similar to the one developed by Francis, and to his invention of a simple device for measuring a machine's power output, a modified form of Prony's original dynamometer. Details of one of his several patented devices are shown in Figure 2.27.

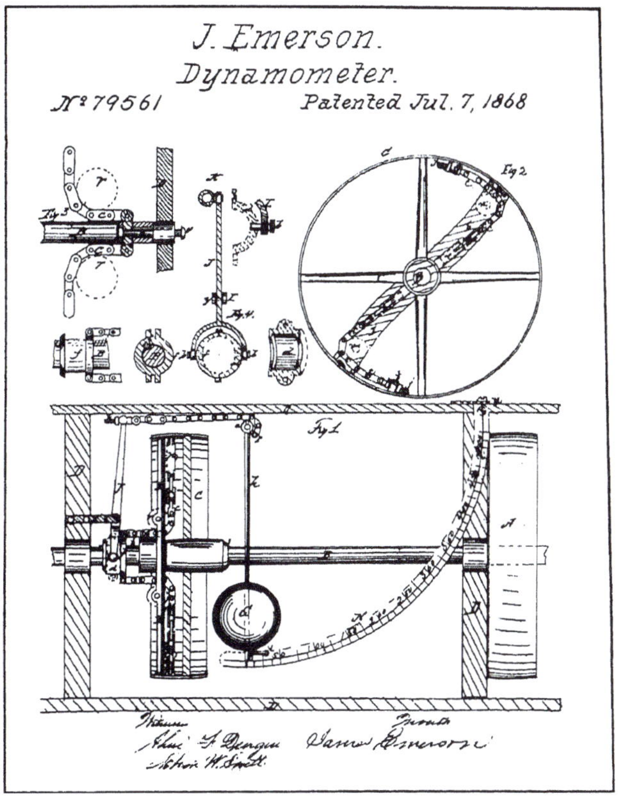

Figure 2.27 Patent drawings of Emerson's dynamometer, 1868

Emerson's dynamometers enabled him to reduce the cost of testing 30- to 50-fold, and he was therefore able to set his fees at a level that even the smallest turbine builder could afford. His invention also enabled tests to be carried out quickly and cheaply *in situ* by a semi-skilled technician, and thus represented a major breakthrough in turbine testing. For the turbine builder, it provided valuable information to supplement the results of model tests.

3.7 The coming of hydroelectricity

In this section, I want to provide a review of the impact that the development of hydroelectric power generation has had on turbine development. The period covers some fifty years, and the treatment is, for reasons of space, necessarily compressed. The development of practical methods of generating electricity in the 1870s, and the subsequent introduction of electric lighting, saved the water turbine from almost certain oblivion as a major industrial prime mover. Water power was already being eased out of many of its traditional markets by steam, which had itself undergone revolutionary improvement since the early work of Newcomen and Watt. The steam engine had become an economical source of mechanical energy. It could operate all year round, providing flexible and secure supplies of power. What is more, steam could be used almost anywhere that fuel was available. It was not affected by flood or drought.

The clamour for public electricity created a new market for the water turbine for which the US turbine manufacturers were singularly ill prepared. For general industrial use, their machines could not be bettered; they were cheap, efficient and reliable. But they had evolved empirically rather than by a process of rational design, whereby changes are suggested by theoretical analysis of the behaviour of the machine. With only limited understanding of hydraulic science, US manufacturers could not easily adapt their designs to meet the new and stringent performance characteristics demanded. Indeed, analysing the complex flow patterns through such machines would have defied even the most competent of nineteenth-century mathematicians.

The challenge for turbine manufacturers was this. Hydraulic motors extracting several kilowatts from a water flow formed their standard product, for which once-high sales were falling off. A potential market was emerging for a smaller number of one-off units designed to extract several megawatts. There were opportunities for those who knew how to scale-up by a factor of one thousand, but problems for those who did not.

It was hardly surprising, therefore, that the US power companies turned to Europe for their hardware. European manufacturers had maintained the mathematical traditions established by their founding fathers (the Eulers, Borda, Poncelet and others) and continued to advance hydraulic theory.

Swiss and German manufacturers, in particular, were accustomed to designing turbines from first principles to suit the wide range of conditions that existed in their native countries. It was less difficult for them to adapt to the demands of the power companies.

One such company, the Cataract Construction Company, blazed the trail into the new era of hydroelectricity by proposing, and later building, a scheme to harness the vast river flow over the Niagara Falls. The first stage of its ambitious scheme was completed in 1896 and involved the construction of three giant turbines each capable of turning at 250 rpm under the head available (over 40 m) and generating 3.7 MW (5000 horsepower). There was nothing like this in the storerooms of US manufacturers, nor did they have experience with casting components on such a scale.

Predictably, the tender for these turbines went, not to a US company, but to a Geneva-based firm. And what is more, the turbines chosen were of an outward-flow design based on those of Fourneyron, which were still widely used in Europe. Although less efficient than contemporary mixed-flow turbines, the Swiss machine was judged to be a safer proposition given the company's experience with manufacturing large turbines and the unprecedented requirements of power, speed and head.

The Niagara scheme was not the first hydroelectric system to be built. That distinction fell to Sir William Armstrong who lit his Northumberland home (Cragside, now a UK National Trust property) as early as 1880, by harnessing a small stream that flowed through his garden. However, Niagara was by far the biggest. It was, above all, a showpiece of engineering and one that stimulated major interest in turbine development.

Before long, the attention of European manufacturers turned to improving the American mixed-flow turbine because of its superior mechanical design and greater efficiency. The immediate objective was to increase the speed and power of the wheel and adapt it for operation at much higher heads.

Exercise 2.4

Think about why hydroelectric turbines are being described as needing to work with high heads and to turn at high speeds.

It was during this period, in the closing years of the nineteenth century, that the familiar outline of the modern Francis turbine finally took shape, more than fifty years after James Francis had tested his first model centre-vent machine – and over 4000 miles away. The turbine was fundamentally redesigned along scientific principles. The performance of each component was analysed and tested and, where appropriate, modifications were introduced.

Ancillary equipment was also evaluated. Governing systems, in particular, which regulated the flow to maintain the rotor at constant speed, were much improved by research. Constant speed was an absolute requirement for electricity generation.

Other advances included:

- the rediscovery of the flared draft tube, made possible by the introduction of reinforced concrete;
- the development of more effective scroll cases and guide vane arrangements which imparted the correct amount of swirl to the water so that it entered the runner without shock;
- pivot gates which permitted greater control over the volume flow than had been possible with sliding cylindrical gates or even large butterfly valves.

These were not new ideas; they had been tried before. For example, the first scroll case was patented as early as 1829 by Zebulon and Austin Parker, the inventors of the draft tube. It was one of those inventions that appear to have been conceived as a result of an accident. A plank had fallen into the water whilst the two brothers had been testing one of their early horizontal wheels. It became wedged across the intake causing the water to swirl about the runner and the power output to increase dramatically.

By the early years of the twentieth century, US turbine manufacturers were beginning to make a comeback, and by the 1920s they had regained their engineering supremacy. The best machines were being designed in US workshops by a new generation of US engineers, who combined familiarity with the latest hydraulic theory with the intuitive craft skills and entrepreneurial flair of their predecessors.

In the years that followed, larger and larger Francis turbines were successfully developed in the USA and elsewhere. By the 1930s, several machines with power output in the region of 20 MW and rotating parts weighing in excess of 200 tonnes were in service. Even larger machines have since been built. Some today have ten times this capacity, generating over 250 MW with a conversion efficiency greater than 90%.

4 REVIEW OF TURBINE DESIGN

4.1 Mechanics: theory and experiment

There are two distinct approaches to product development. The first involves the use of theory to suggest ways in which a device or mechanism might be improved. The second relies almost entirely on experimental techniques and craft skills. To be successful, both approaches depend on intuition and prior knowledge and experience in the field.

The theoretical approach is exemplified by the work of Jean Victor Poncelet and Benoît Fourneyron in the 1820s and 1830s. Poncelet used the theories of Borda and his contemporaries to redesign the traditional undershot water wheel, and doubled its performance as a result. He used a deep-angled sluice and curved vanes to reduce shock as the water entered the wheel, and ensured that virtually all of the kinetic energy had been removed before the water was discharged into the tail race. Fourneyron adopted a similar approach a few years later when designing his first water turbines. He took special care over the shape of the runner and guide vanes.

The foundations for the experimental approach to design were laid by John Smeaton, Antoine de Parcieux and others in the 1750s. Their technique involved making minor alterations to individual components of the device, or adjusting its operating conditions, and assessing the effect that these had on overall performance. Beneficial changes were then selected and incorporated in new designs.

As we have seen, the technique was used with spectacular success in North America in the second half of the nineteenth century to improve the performance of the basic Francis turbine. Despite their ignorance of, and indifference to, theoretical considerations, a generation of talented small-time engineers and turbine builders confounded theoreticians by developing progressively cheaper, more reliable and ever more efficient machines.

In reality, the two approaches to design are complementary: each has its strengths and its limitations. Theory on its own can only go so far towards suggesting how improvements might be made. It invariably relies on approximations and oversimplifications that frequently break down when applied to highly complex systems. For example, the hydraulic theory available to early pioneers like Poncelet and Fourneyron was necessarily crude and often failed to predict or explain experimental results. Poncelet himself tested his theoretical ideas on scale models and determined experimentally the correct ratio of wheel-to-water speed. Even today, there are many areas of engineering where theory is unable to predict precisely how a machine, mechanism or structure will behave in the real world and where other approaches to design must be sought.

What is more, theory can actually delay progress. This has often happened when poor or erroneous theory has been used to justify particular ideas or perpetuate bad practices. Bad theory, unlike bad practice, is seldom self-evident. It may take many years to unmask theoretical fallacies, and, conversely, academic zeal may sometimes be responsible for stifling novel and original approaches that would otherwise lead to improvement.

It has been suggested that the highly successful development of the American mixed-flow turbine would not have happened in the way it did had more attention been paid to theoretical considerations.

However, the empirical, 'cut-and-try' approach also has its limitations. The procedure is often time consuming and costly and involves tedious repetition. But most importantly, it lacks direction and generally results in uncertain and unsteady progress. Such problems were very apparent in the USA in the 1890s, when turbine manufacturers were unable to adapt their empirically designed machines to meet the demands of the new power companies. They had insufficient understanding of hydraulic theory to be able to design turbines to close specifications.

The success of the experimental approach may also depend as much on luck as intuition. It was luck that led the Parker brothers to the idea of a scroll case.

Thus the most effective design policies usually embrace both theory and practice. The former provides guidance and suggests new concepts – or indicates where fundamental physical limitations exist, while the latter corroborates the theory and fills in gaps, ultimately providing a way forward when theory breaks down or cannot readily be applied.

4.2 Materials and manufacture

Two other factors which can have a dramatic effect on design are both technological developments – the use of new materials and the development of new manufacturing techniques. The discussion of water wheels well illustrates the impact that new materials can have on design. The introduction of cast, and later wrought, iron enabled larger and more powerful machines to be built, and turned a relatively inefficient and unreliable energy converter into an indispensable source of power which played a key role in the Industrial Revolution.

Millwrights began to make use of cast iron in the closing years of the eighteenth century. It was used initially to replace heavy wooden axles, but it was not long before complete water wheels were being fabricated. These wheels were considerably stronger, longer lived and less prone to breakdown than the wooden wheels they replaced. Cast iron was less subject to wear: it did not decay, nor did it shrink or warp.

But cast iron did not reign for long. It was soon superseded by an even better material, wrought iron. Wrought iron was physically very different from cast iron, which was hard and brittle and could neither be worked, except by melting and recasting, nor used under tension, which caused it to fracture. (The physical structure and properties of cast iron have been considerably modified by metallurgists since the eighteenth and nineteenth centuries, and so cast iron today is unlike that produced originally.)

The introduction of wrought iron made possible new, more complex shapes and encouraged innovative designs. These included curved vanes and buckets which reduced turbulence and helped to improve conversion efficiencies, and rim power take-off and tensioned spokes, which enabled much lighter axles to be used and allowed power-to-weight ratios to be substantially increased.

The switch from wood to iron inevitably involved changes in manufacturing techniques too. Blast furnaces and foundries took on increasing importance, whilst the traditional millwrighting profession declined.

SAQ 2.2 (Outcome 2.1)

The development of water turbines has not happened at a steady pace. At times progress has slowed almost to a halt, whereas at others major improvements have been made in quick succession. Suggest reasons for this, and support your argument with examples from the text.

4.3 Who invented the Francis turbine?

We are now in a position to trace the evolution of the Francis turbine and to assess the role played by Francis. I should like to begin by considering the key component of the turbine – the runner – and then make a few observations about its setting.

Figure 2.28 shows the ancestry of the modern Francis turbine in diagrammatic form, indicating how the significant advances led on from one another.

The idea of channelling the water through a horizontal wheel to a central discharge was first suggested by Jean Victor Poncelet in 1826 as a means of overcoming two fundamental limitations of his rationally designed undershot water wheel. These were, first, that water was only admitted to a part of the wheel rather than around its entire circumference and, second, that its flow was reversed inside each bucket since the water entered and left via the same opening. Flow reversal meant energy being wastefully dissipated in turbulence. Poncelet recognized that with full circumferential admission more water would pass through the wheel in a given time, enabling an equivalent output to be produced by much smaller – and hence cheaper – wheels.

However, Poncelet did not conceive the principle of the vortex chamber, which caused water surrounding the wheel to rotate at such speed that it entered the wheel without shock. This was the essence of the device patented by Samuel Howd in 1838.

Unfortunately, little is known about Howd and his 'Improved Water Wheel' apart from the description and drawings in his patent. It is not clear, for example, whether Howd was aware of Poncelet's ideas when designing his inward-flow wheel. It seems probable that he saw his invention as a development of the traditional flutter wheel which was widely used in the American backwoods at the time.

What is clear is that any claim that Howd might have made to have been one of the first to recognize the importance of the radial inward-flow configuration must be tempered by the fact that he later rejected the concept outright. His 'Latest lmproved Water Wheel', patented in 1842, was designed around an outward-flow wheel. Again, whether Howd got the inspiration for his new machine from others is not known, although it is interesting to speculate on this. Ellwood Morris's descriptions of Fourneyron's outward-flow turbines began to appear in the USA the very same year.

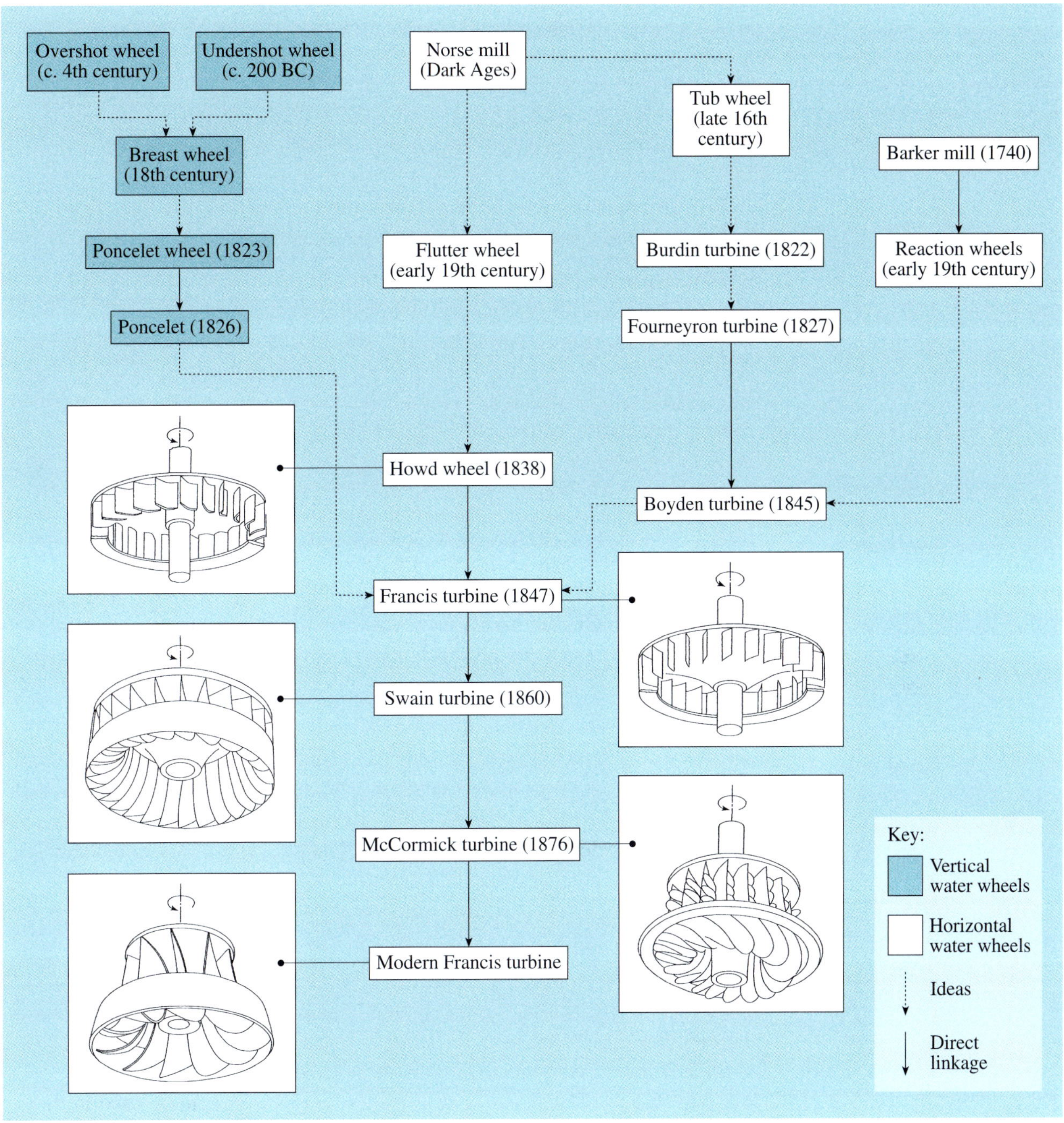

Figure 2.28 The evolution of the Francis turbine

However, there was no such doubt in Francis's mind when, some years later, he recognized the potential of Howd's original turbine, and its centre-vent configuration in particular, and selected it in preference to the beautifully engineered outward-flow machines of Boyden that were popular at the time. This must have required considerable courage on Francis's part, since he was publicly rejecting the conventional wisdom of the day which held that the 'centrifugal' force of the outward-flow turbine augmented the discharge and increased the power of the wheel. By this logic, the centre-vent principle was inferior.

With his considerable grasp of hydraulic theory and the practicalities of turbine operation, Francis redesigned Howd's machine along more scientific lines. Assisted by skilled mechanics, and with Lowell's excellent workshop facilities at his disposal, he produced a much larger capacity, more robust machine with improved performance characteristics which was intended for use by the Lowell textile mills. The finished product bore little resemblance to Howd's crude original, although, according to Francis, it would in all probability have been covered by Howd's patents.

Unfortunately for Francis, very few of his machines were built. They appeared to offer no significant advantages over the well-established Boyden–Fourneyron turbines. But the concept of a wheel with a peripheral feed and a central discharge did give rise to the highly successful American mixed-flow turbine which, as we have seen, evolved over the following half century and largely as a result of the efforts of an army of small-time millwrights working without knowledge of, or interest in, theoretical concepts.

I have discussed the contributions made by two of the more prominent millwrights in this period: Asa Swain, who introduced deep buckets into his wheels to increase the capacity, and John McCormick, who developed the incredible 'Hercules' turbine, which was the first of a new generation of more powerful, faster-turning machines.

Swain claims to have invented the concept of a mixed-flow runner, although the first radial-flow machine to incorporate axial flow was actually constructed by Uriah Boyden in 1849, a decade before Swain's first turbine. The idea probably came from Francis – at least that is what he later claimed. Of course, axial flow itself was nothing new: it had been used several centuries earlier in horizontal wheels that evolved from the primitive Norse mill.

Let me note in passing that there were several external factors that helped to make this extraordinary period in the turbine's evolution so successful. One of these was the availability of the Prony friction brake, which enabled a turbine's efficiency to be accurately and conveniently measured; another was the introduction of commercial testing, which was of considerable assistance to the genuine turbine manufacturers in helping them to improve their products. Some have argued that the latter was as necessary to the development of the reaction turbine as the whole 'cut-and-try' period. One without the other would have been useless.

Testing also fulfilled an equally important function of precipitating the demise of many half-baked and crackpot ideas. Much of the credit for this goes to the eccentric Mr Emerson, who recognized the need for such a service and duly provided it.

By the early years of the twentieth century, the mixed-flow runner – and the ancillary equipment to go with it: the scroll casing, pivot guide vanes, draft tube, etc. – had evolved into broadly its current form. It was the product not of one mind but of many, as can be seen at a glance from Table 2.4, which records some of the more important inventions which have helped to make this type of turbine such a successful prime mover.

Table 2.4 Evolution of the Francis water turbine: some key developments

Wheel		Wheel setting	
1826	Poncelet suggests inward-flow principle	1829	Scroll case (Parker brothers)
1838	Howd's 'Improved Water Wheel': full circumferential feed with reaction and guides	1833	Draft tube, without flare (Parker brothers)
1847	Prototype Howd–Francis turbine	1847	Diffuser (Boyden)
1849	Francis's Boott mill turbine	c. 1850	Iron scroll without guides
1849	Boyden's mixed-flow turbine	1854	Wicket gates (Roberts)
1858	Swain's mixed-flow turbine with deep buckets giving increased capacity	1860	Iron scroll with guides (Warren)
1873	The Obenchains' 'Little Giant'	c. 1870	Thomson's 'Vortex' turbine with scroll and pivot guides
1876	McCormick's 'Hercules', increased output and higher speed	c. 1915	Pivot gates (Finche)

Source: adapted from Safford and Hamilton (1922)

Of course, a complete table would need to include reference to many more individuals, especially if it was to trace the origin of all the ideas incorporated in the machine right back to the early water wheels. It would include such characters as Leonardo da Vinci, John Smeaton and Benoît Fourneyron, as well as the army of mathematicians, physicists and engineers who built up the theory and science of hydraulics.

How, then, did this machine come to be called the 'Francis' turbine? It is a good question, and one that has often been asked over the years. Some have suggested that it should rightly be called the 'Howd' turbine, or the 'Howd–Francis–Swain–McCormick' turbine. Indeed, the modern Francis turbine, which is composed of many parts, probably bears less resemblance to the original Francis turbine than that machine did to Howd's 'Improved Water Wheel'. An alternative designation would be the 'American' turbine, the name by which it was known in the USA for many years – it is, after all, an almost wholly US invention.

The Europeans, it appears, are responsible for the name. Although Francis's original centre-vent machine was a commercial flop in his native country it did see service in Europe, where many simple Francis turbines were built in the latter half of the nineteenth century. And when, in the 1890s, European manufacturers were attempting to develop higher-performance machines to drive large generators in hydroelectric schemes it was to the Francis turbine that they turned. The basic Francis turbine offered the prospect of better efficiency and higher rotational speed. It was this that finally killed off the Fourneyron machine. The name 'Francis' has stuck, and is today almost universally applied to what is really an American mixed-flow turbine.

In conclusion, we can see that although Francis played an important role in the development of what has come to be called the Francis turbine, he did not conceive the principle upon which it works; neither did he invent much of the hardware that goes with it. Like Smeaton and de Parcieux before him, Francis was a designer and improver rather than an inventor.

Francis adopted a scientific approach to solving technical problems of turbine design and perfecting mechanical construction. He was familiar with the hydraulic theory of the day and was guided by it in this work. He was also a gifted experimentalist, although he relied greatly on the engineering skills of those around him and especially on those of his colleague, Uriah Boyden. It should not be forgotten that it was Boyden who supplied much of the data on which Francis worked and who pioneered many of the experimental techniques which he so successfully applied.

Francis was unquestionably a great man who contributed much to the furtherance of US hydraulic science. But he is remembered more for his contributions to large-scale hydraulic engineering rather than turbine design. Louis Hunter has observed that when two committees of leading engineers reviewed Francis's career shortly after his death in 1892 they gave only incidental and slight attention to the turbine that bears his name.

SAQ 2.3 (Outcome 2.2)

Give some examples of how the arrival of new materials and manufacturing methods contributed to the development of water power, from Vitruvius's water wheel to the Francis turbine.

5 ESSENTIAL TURBINE ANALYSIS – FROM FLOW TO ROTATION

In this section, we will look at two ways in which energy stored by water in a reservoir can be converted into the rotational energy and work done by a machine. Figure 2.29 illustrates the two different pathways. In both cases a water turbine performs the crucial transfer of energy from a flow into the rotational energy of a wheel (or rotor or runner). The transfer of energy from the water to the turbine happens when blades on the rotor are put in the path of the flow. This energy transfer will occur only if the water is slowed down or caused to change its direction of motion by the blades.

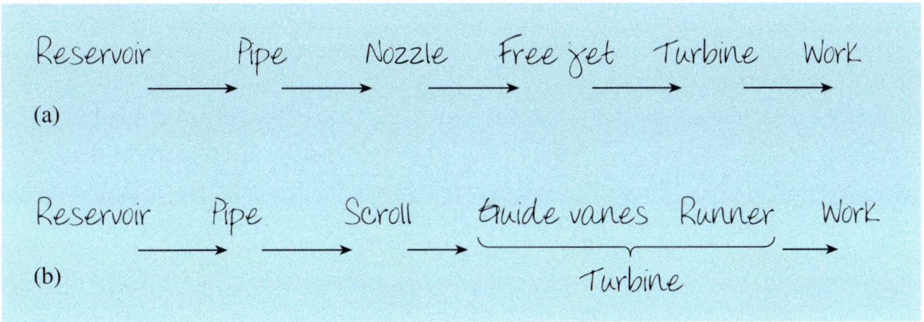

Figure 2.29 Pathways for energy transfer for (a) a Pelton Wheel and (b) a Francis turbine

As you have seen in the case of the hydroelectric electricity generation (Part 1 of this block), some turbines can also be used as pumps. In a pump, the rotor is driven and its rotational energy is transferred to flow energy.

I will consider two types of water turbine: a Pelton wheel (an example of an impulse turbine) and a Francis (reaction) turbine, corresponding respectively with the pathways illustrated in Figure 2.29(a) and (b). Only the second of these can be reversed to make a pump.

I will start by examining the transfer of energy in and around a Pelton wheel.

5.1 Fluid flow and energy around a Pelton wheel

The principle of the Pelton wheel is shown in Figure 2.30. The water is channelled down a pipe starting at a high point and ending lower down. The water leaves the pipe through a nozzle and hits the blades of the turbine at high speed to make the wheel turn.

Since the turbine is extracting the kinetic energy of the flow by momentum transfer from the water, it is worth knowing what rate of supply of momentum is available from a given head of water. Let's find out the absolute maximum possible speed we could get from a given head of water.

To keep frictional losses small, it pays to use **▼Large pipes with slow flows▲** wherever possible. In the extreme, the pipe would be the same diameter as the reservoir, and the nozzle feeding the water onto the turbine would be just a hole, as shown in Figure 2.30(c).

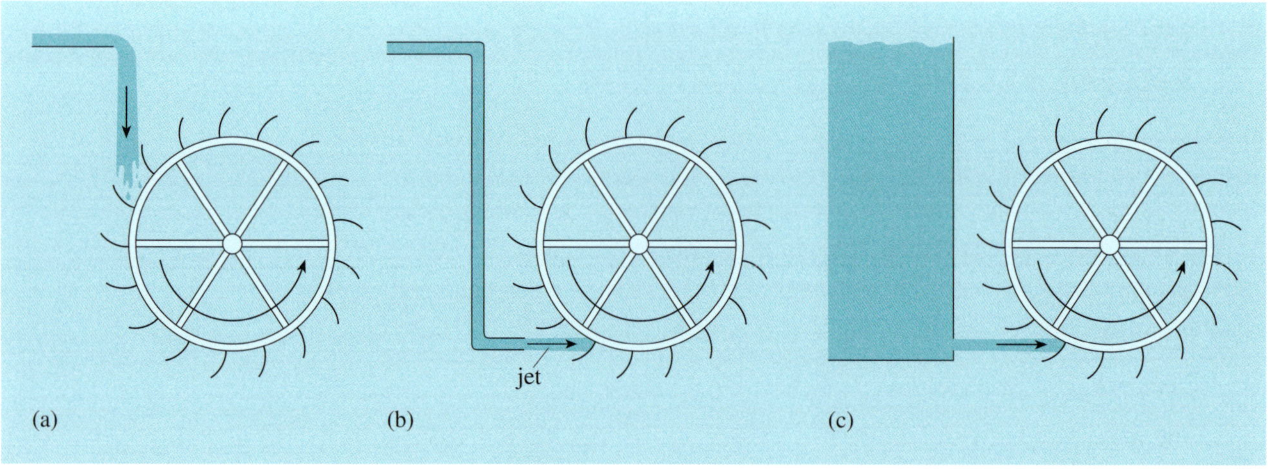

Figure 2.30 Water flow to Pelton wheel: (a) free flow as a waterfall, (b) directed flow down a pipe and through a nozzle and (c) design of reservoir and nozzle for minimum flow losses

▼Large pipes with slow flows▲

A flow in a pipe or channel will lose energy all the time through friction with whatever surface is confining it. As with aerodynamic drag, such energy losses will increase with channel wall area and with the square of the flow speed; in turn, the power loss increases with the cube of the flow speed. For a pipe or channel of circular section, we can say that power losses scale as follows:

$$P \propto 2\pi r l v^3 \quad (\text{J s}^{-1})$$

Similarly, the volume flow through a pipe or channel of circular section scales as:

$$Q \propto \pi r^2 v \quad (\text{m}^3\ \text{s}^{-1})$$

So the energy lost per cubic metre of water delivered down a circular section pipe or channel will scale as:

$$\frac{2\pi r l v^3}{\pi r^2 v} \propto \frac{l v^2}{r} \quad (\text{J m}^{-3})$$

It is no surprise to find that water is carried from the reservoir to the turbine at slow speed in large-radius pipes or channels.

We can use Bernoulli's equation (which is essentially an expression of the law of conservation of energy) to find out the speed of such a water jet. Bernoulli's equation neglects losses arising from viscosity and states that, no matter where you look in a continuous body of fluid, the total amount of energy per unit volume (the energy density) is the same. This energy per unit volume is composed of three elements: the internal energy of the fluid due to the pressure, the gravitational potential energy and the kinetic energy due to the speed of flow. As you move from place to place in the fluid, energy that was in one form may be transferred to another, but the sum of all three terms is always the same.

Bernoulli's equation may be written thus:

$$p + \rho g h + \tfrac{1}{2}\rho v^2 = \text{constant} = E_\text{v} \tag{2.1}$$

where ρ is the density of the fluid and E_v is the energy per unit volume.

The reservoir

If we apply Equation (2.1) to a situation where there is no flow, such as the reservoir, we can say that on the top surface $p = 0$ (if we measure p relative to atmospheric pressure) and, since v is also zero, we can see that the only term in the equation with a finite value is the gravitational potential term:

$$\rho g h_0 = E_\text{v} \tag{2.2}$$

h_0 can be measured from any arbitrary reference height, but for convenience in this instance we will use the bottom of the reservoir as our zero. So on the surface E is just the gravitational potential energy of the water.

At the bottom of the reservoir, $h = 0$, and, since v is still zero, we have:

$$p_\text{btm} = E_\text{v} \tag{2.3}$$

Together equations (2.2) and (2.3) lead to the conclusion that p at the bottom of the reservoir is equal to $\rho g h$ at the surface.

The nozzle

We are supposing that pipework is sufficiently large that it just adds to the effective volume of the reservoir, without incurring significant loss. It is as if the reservoir has a small hole at the bottom that acts as a nozzle. Water is accelerated through the hole, trading pressure energy from the reservoir for kinetic energy in the jet. The water spurting away from the hole has a speed v_jet and is now at atmospheric pressure (since it is exposed to the atmosphere, the pressure per unit volume is as it was at the top of the reservoir, namely zero). The gravitational potential energy is zero as this is where we chose to measure height. So now:

$$\tfrac{1}{2}\rho {v_\text{jet}}^2 = E_\text{v} \tag{2.4}$$

All the energy in the water is now in the form of kinetic energy.

Equations (2.2) and (2.4) relate v_jet to the most measurable quantity for our reservoir, namely the height of its surface above the turbine, and we have:

$$v_\text{jet} = \sqrt{2gh_0}$$

That's interesting. This is the same answer as you would get for the speed of an object falling a distance h_0 under gravity (using the familiar equation from Block 3 Part 1 of $v^2 = u^2 + 2as$, with the initial speed $u = 0$ and gravity providing the acceleration and the distance travelled $s = h_0$). So, v_jet is the speed the water would have attained if it had just been free falling as in a waterfall (ignoring the effect of wind resistance).

Compare this with Section 2.3 and Torricelli's results.

In practice, wind resistance rapidly limits the speed the water gets to in a waterfall – to minimize these drag losses, the water is instead guided to a nozzle down a large pipe.

As the flow speed is fixed only by the pressure head, h_0, the other result we get from this analysis is that the other factor affecting how much water is passed through the nozzle per second is the nozzle's effective area A_n.

In general, a flow of fluid having density ρ and speed v, crossing a perpendicular area A, transfers mass at a rate, in kg s^{-1}, of:

$$\dot{m} = \rho v A$$

The equivalent volume flow rate in m^3 s^{-1} is:

$$Q = vA$$

The means that, in the jet, the mass flow is given by:

$$\begin{aligned}\dot{m} &= \rho A_n v_{jet} \\ &= \rho A_n \sqrt{2gh_0}\end{aligned}$$

Free jet

Now we have expressions for how much water is being delivered per second, and how fast the jet is moving. The next step is to analyse the interaction of the jet with the Pelton wheel. The purpose of the analysis is to be able to design the nozzle and wheel to meet the requirements of a machine or generator to which it is connected.

SAQ 2.4 (Outcome 2.3)

A high-pressure jet-washer from a hardware store pressurizes the water to 100 bar (1 bar of pressure is equivalent to a 10 m head). Estimate the speed of the water jet from the pressure washer.

The turbine

The Pelton wheel transforms flow to rotation by the jet of water hitting the blades of the wheel.

Suppose that the jet of water is initially deflected away so that it does not hit the blades of the wheel, which is stationary. It is not yet connected to its load, but is free to turn. The jet of water is then directed to hit the blades on the rotor. The blades are pushed forward and the wheel starts to rotate. As the jet of water continues to hit the blades the wheel will speed up. Because it can turn freely the impact of the jet will accelerate the wheel. The difference between its speed and that of the water in the jet will diminish as it accelerates – the blades are effectively running away from the jet. The pushing effect of the jet on the blades will decrease until it just balances any friction in the rotation of the wheel. Now if some load is connected to the wheel, the wheel will be slowed down whereupon the jet would have a greater effect. Thus a balance will be achieved where the wheel has slowed enough for the pushing effect of the jet to balance the load.

You should notice one thing about this informal analysis. The pushing effect of the jet depends on the relative speeds of the jet and the blades or buckets. If the blade and the jet were moving at the same speed then the jet would not hit the blade and thus not exert any pushing effect. I want to see how much power can be extracted from a Pelton wheel. This power is the rate at which energy is transferred from the jet to the wheel and then on to the connected machinery.

The shape of the blade is carefully designed so that water flows around the blade smoothly. Also none of the jet (or as little as possible) is wasted between the blades. The shape of the blade has been developed by iterations into a double cup shape (see Figure 2.31). This splits the jet in two and guides it around the inside of the cup, leaving in the opposite direction.

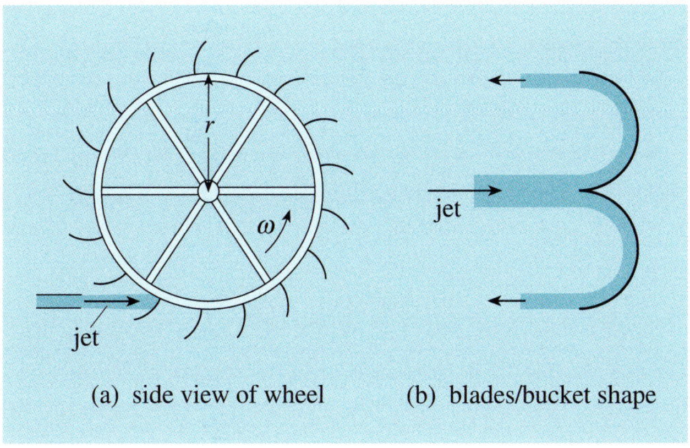

Figure 2.31 Pelton wheel: (a) side view of wheel and (b) blade shape

5.2 Torque and power for the Pelton wheel

We will now examine in more detail how fast the wheel spins for a given load or power taken from the wheel. Watch out for the careful use of *velocity* to mean a given speed in a given direction.

Every blade on the wheel goes round at angular speed ω. The jet hits each blade in turn at a radius r metres from the centre of the wheel. The tangential speed of a blade is therefore $r\omega$. The speed of the jet is v_{jet}, so the jet hits the blade at relative *velocity* $v_{jet} - r\omega$, in the 'forward' direction. The jet flows round the surface of the blade until it leaves with a relative velocity $-(v_{jet} - r\omega)$, except for some losses caused by friction with the blade. However, for a well-designed blade these are small. The negative sign indicates that this velocity is in the opposite direction to the incident velocity of the jet.

Let us look more closely at what is happening in one single blade (see Figure 2.32).

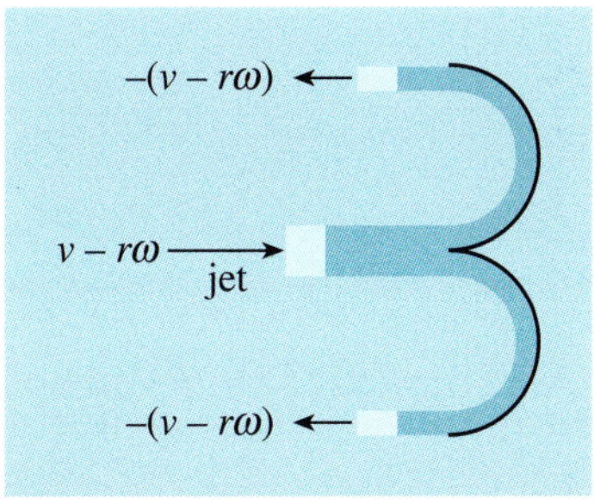

Figure 2.32 Jet entering and leaving cup

The relative velocity of water entering a *single* blade is:

$$(v_{jet} - r\omega)$$

Mass leaves with relative velocity:

$$-(v_{jet} - r\omega)$$

Its velocity has therefore changed by:

$$2(v_{jet} - r\omega)$$

The water that does not hit the one blade will be hitting the next blade on the wheel. So the total rate at which mass interacts with the blades is $\rho\ v_{jet}\ A$

and it has its velocity changed by $2(v_{jet} - r\omega)$. Thus the rate at which momentum is being changed is given by:

$$2(v_{jet} - r\omega) \times \rho v_{jet} A$$

Newton's Second Law says that a rate of change of momentum equals a force.

To change the momentum of the water requires a force exerted by the blade. By Newton's Third Law, there is an equal and opposite reaction force on the blade which turns the wheel. The force in newtons on the wheel is thus:

$$F = 2(v_{jet} - r\omega)\rho v_{jet} A$$

This force gives a torque Γ to turn the wheel:

$$\Gamma = Fr = 2(v_{jet} - r\omega)\rho v_{jet} Ar \qquad (2.5)$$

The power in watts produced by this torque is:

$$P = \Gamma\omega = 2(v_{jet} - r\omega)\rho v_{jet} Ar\omega \qquad (2.6)$$

The jet constantly changes momentum (by changing direction in flowing round the cups in the blades), giving a continuous force to the blades on the wheel. The resulting torque accelerates the wheel. If there is no load on the wheel, such as a machine or generator, then the wheel spins up to full speed ω_{max}, at which speed the jet exerts no force on the wheel. If no force is exerted, there can be no energy transfer from the jet to the wheel. If a load is now attached to the wheel it will slow down. At what angular speed will it settle?

If machinery attached to the wheel requires a torque, Γ, then the wheel will slow down to an angular speed given by Equation (2.5). Plotting a graph of Γ against ω shows a straight line (see Figure 2.33).

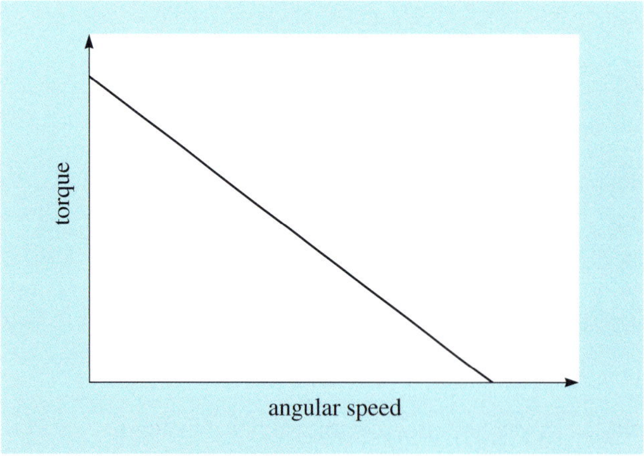

Figure 2.33 Graph of torque against angular speed for a Pelton wheel

The power, P, delivered by the turbine at this angular speed is given by Equation (2.6). Plotting a graph of P against ω shows an inverted U-shape (see Figure 2.34).

This graph is interesting because it shows that there is a particular angular speed at the top of the curve where the power is a maximum. At this speed, the turbine is transferring as much energy as possible from the flow in the jet to the rotating wheel (and anything that is connected to it). That is where an energy-conscious user would like to operate the turbine.

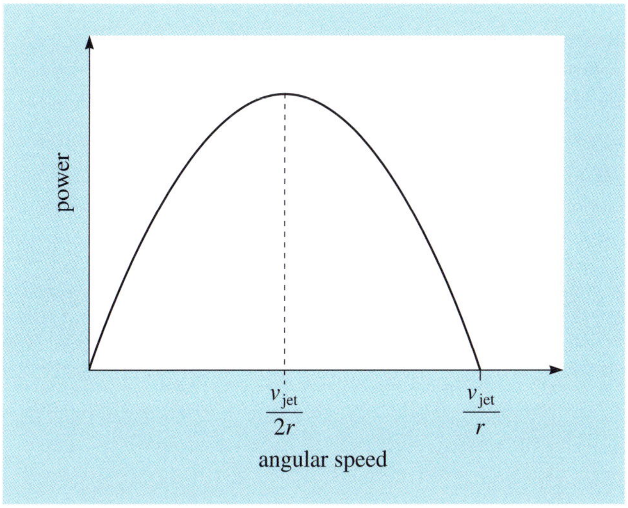

Figure 2.34 Graph of power against angular speed for a Pelton wheel

The other interesting thing about this graph is that the maximum power is extracted when the tangential speed of the wheel, $r\omega$, is exactly half of the water jet speed, v_{jet}. Looking back at the analysis of the turbine, we see that the water leaving the blade has a velocity relative to the blade of $-(v_{jet} - r\omega)$ in the forward direction. But $r\omega$ is $v_{jet}/2$, so this relative velocity is just $-v_{jet}/2$. Now, the blade is moving relative to the ground at a velocity of $v_{jet}/2$, so the horizontal velocity of the water leaving the blade relative to the ground is $v_{jet}/2 - v_{jet}/2 = 0$. This should not surprise us. It just shows that all the kinetic energy, and hence the maximum power, is being extracted from the flowing water.

The equations and associated graphs of torque and power can help an engineer to predict how a particular turbine will behave under different types of load. They also allow engineers to design a turbine for particular circumstances, by altering the radius of the wheel, the number and the cross-sectional area of the jets, and the gearing between the wheel and the load.

SAQ 2.5 (Outcome 2.3)

A Pelton wheel has a radius r of 2 m, and the water jet's cross-section and speed are 1×10^{-3} m^2 and 30 m s^{-1} respectively. The density of water is 1000 kg m^{-3}. Determine the following:

(a) the maximum torque the system is capable of providing;

(b) the angular speed of the wheel under no-load conditions;

(c) the maximum power the wheel can produce;

(d) the efficiency at which the wheel will operate when the load imposes a torque of 2200 N m.

5.3 The Francis turbine

In this section, we will examine an inward-flow Francis turbine, designed to operate at a particular angular speed. I have in mind something like the turbines used at the Dinorwig hydroelectric pumped-storage facility. To contribute a.c. electricity to a national grid, the rotation of any generator will be synchronized with the entire system and that means that it will have to run at a specific speed.

The most immediately apparent difference between the impulse turbine and the Francis turbine is that the latter is immersed in the flow of water from the reservoir. Nozzles and jets are not used and it works on different mechanical principles. Another difference between the turbines that has enormous practical significance is that the reaction turbine, being fully immersed, is able to act as a pump if driven in reverse. This is one reason why the Francis and not the Pelton type was used at Dinorwig.

The Francis turbine is said to be an example of a *reaction* turbine. In the Pelton wheel, the water is already at atmospheric pressure when it hits the wheel and all its energy is kinetic. In a reaction turbine, pressurized water enters the turbine and the pressure in the water falls as it passes through the runner. At the inlet, therefore, the energy in the water is partly kinetic and partly hydrostatic (pressure). As the water passes through, the energy stored within it in the form of pressure is steadily transferred into additional kinetic energy in the flow, so that by the time the water emerges at the output of the turbine there is no pressure head left, and most of the kinetic energy has been transferred into the rotational kinetic energy of the turbine runner and work done by the machine to which its shaft is connected.

We can analyse the energy transfer in a Francis turbine by comparing the momentum of the water on entry to and exit from the turbine, as we did for the Pelton wheel. As an example, let us look at the type of inward flow turbine used in the Dinorwig hydroelectric power station (see Figure 2.35).

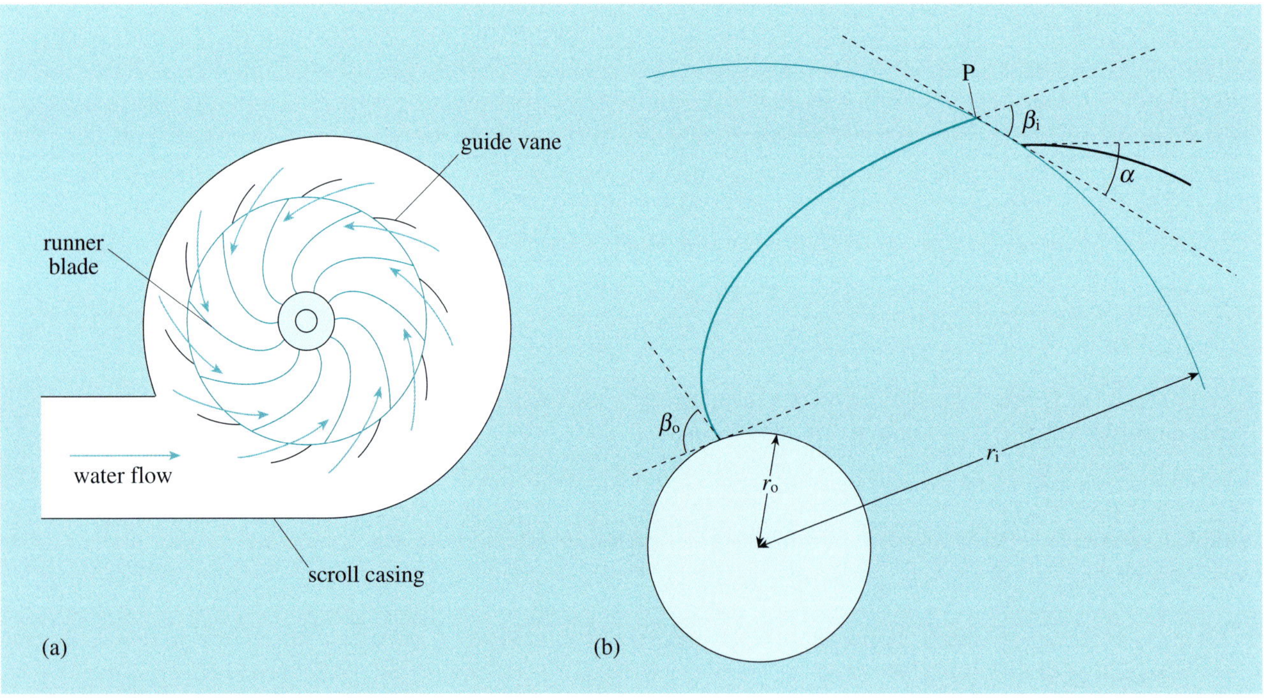

Figure 2.35 Francis turbine – shape of the blades (simplified): (a) general view of runner and guide vane arrangement and (b) detail showing definitions of angles used in analysis

The reservoir and pipework

The flow is pressurized by the gravitational potential energy of the water stored in the reservoir. In the pipe that leads to the turbine some of the stored potential energy (gravitational and pressure) is used to accelerate the fluid, giving it kinetic energy.

The scroll and guide vanes

Water is directed round the turbine by the scroll and then it passes through guide vanes set at an appropriate angle into the runner. As with the Pelton wheel, relative velocities are critical to efficiency. The scroll's job is to distribute the flow evenly round the perimeter of the turbine so that full use can be made of its surfaces. The guide vanes deflect flow into the turbine and are adjusted to control the flow rate – this effectively controls the output torque and power, as we shall see.

We will look at how the angle of the guide vanes sets the flow rate and matches the flow onto the turbine blades. To do this, the vanes must guide the flow onto the leading edge of the runner blades, taking account of the rotation rate of the runner. At the exit of the turbine, we will also see how the trailing edge of the runner blades must be angled. The aim of the design that we will investigate is to achieve efficient operation at a specified speed, as is required for hydroelectric power generation. At the design speed, water should rush in, deliver its energy and emerge placidly. Away from the specified speed, we can expect that turbulence from splashing and whirling in and downstream of the turbine will 'waste' significant amounts of energy.

The turbine

Figure 2.35 shows the geometry of the turbine and how the angles we shall be working with are defined.

I will use a subscript 'i' to denote the inlet, and 'o' the outlet of the turbine. For the water, u will denote absolute components of velocity, and v components of velocity relative to the turbine blade. The velocities will be resolved into radial and tangential components with subscripts 'r' and 't' respectively.

Let's begin with the runner already spinning at angular speed ω. The radius of the tip of the blade (the point labelled P on Figure 2.35) is r_i, which makes the tangential speed of the tip of the blade equal to $r_i\omega$.

The water enters the spaces between blades, where it is forced to follow the ducts in the runner to reach the exits near the axis. Since the blades are quite close together, it is safe to assume that the relative flow is the same across each duct, varying only with radial position.

In our analysis, we will need to look at both the absolute velocity of the water and its velocity relative to the runner blade. Comparing the *absolute* velocities before and after the runner allows us to calculate the power extracted by the turbine, and looking at the *relative* velocities allows us to work out the blade and guide vane angles.

So:

$r\omega$ is the blade velocity

u is the absolute water velocity

u_r is the radial component of u

u_t is the tangential component of u

v is the water velocity relative to the blade

and all of these terms will have the subscript 'i' or 'o', depending on whether we are referring to the inlet or the outlet of the turbine.

Figure 2.36 shows how these velocities relate to one another and to the geometry of the turbine.

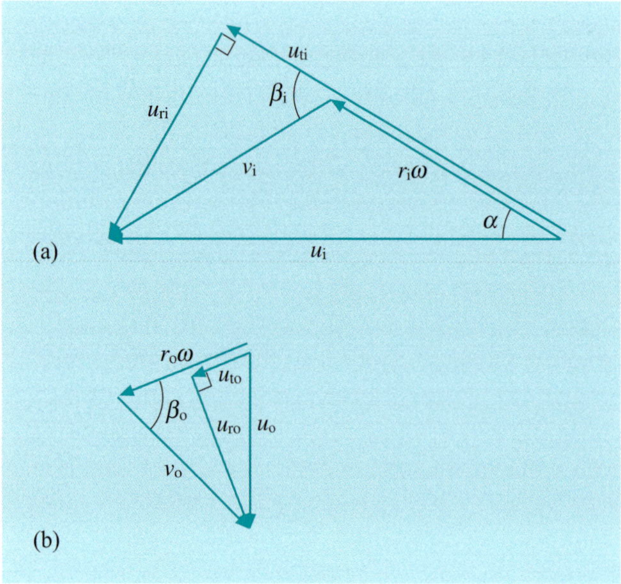

Figure 2.36 Francis turbine – velocity vectors for flows and runner blade (a) at input and (b) at output of turbine

Radial components of inlet and outlet velocities

All the water flowing through the turbine has to cross the cylindrical surface swept out by the runner blade tips. If we assume that this inward flow of water is equal all round the runner, we can say that the radial component of the water's velocity at any point on the cylinder described by the outer tip of the runner is equal to the volume flow rate Q divided by the surface area A_i of this cylinder. A_i is the circumference of this cylinder multiplied by the height Z_i of the runner blades at the tips.

So:

$$u_{ri} = \frac{Q}{A_i} = \frac{Q}{2\pi r_i Z_i}$$

And since what goes in must come out of the turbine:

$$u_{ro} = \frac{Q}{2\pi r_o Z_o}$$

Remember, the runner does not move radially so a component of relative radial velocities and an absolute radial component are identical.

The inlet blade angle for smooth flow

One way to regulate the output power would be to use the guide vanes to also restrict the inlet area, closing it down like a tap. Alternatively, a second set of vanes upstream of the turbine could do this, leaving the main vanes with the important task of steering the flow smoothly onto the runner. Any mismatch will lead to turbulence and hence energy loss.

Looking at the velocity diagram at the inlet (Figure 2.36), we can see that the tangential and radial components of the absolute water velocity are related to each other and the angle α of the guide vanes by the expression:

$$u_{ti} = \frac{u_{ri}}{\tan\alpha} \qquad (2.7)$$

For the water to flow smoothly onto the blade, its velocity relative to the blade at the tip must be parallel to it – we need to take account of the fact that the blades are 'moving targets'.

In the tangential direction, the velocity of the water *relative to the blade* is $u_{ti} - r_i\omega$. The blade has zero velocity in the radial direction, so the radial velocity of the water relative to the blade is just its absolute radial velocity u_{ri}. To get the total relative velocity of the water parallel to the blade, this means that the angle β_i of the blade at inlet must be set so as to satisfy the expression:

$$\tan\beta_i = \frac{u_{ri}}{(u_{ti} - r_i\omega)} \qquad (2.8)$$

Similarly, at the outlet:

$$\tan\beta_o = \frac{u_{ro}}{(r_o\omega - u_{to})}$$

This determines the blade angle at the outlet, provided we can decide what the tangential outlet component of relative velocity should be.

5.4 Torque and power for the Francis turbine

The mechanical principle of a reaction turbine such as the Francis turbine is that a torque is exerted on the turbine blades as a reaction to the changing angular momentum of water passing through the turbine.

Effectively, the torque is applied by the water being turned as it flows through the channel, pressing against the blade as it turns. These forces on the blade all 'add up' to produce a torque on the wheel. But to add them up literally is not easy. Instead, we look at what has happened to the water between entering the channel and leaving it. The *difference* in angular momentum tells us what effect the water has had on the blades as it flows through the rotor.

The angular momentum of the water at the inlet is the momentum tangential to the inlet times the radius at inlet. The rate at which angular momentum enters the turbine is therefore the mass flow times the tangential velocity times the radius. Mass flow is the volume flow Q times the density ρ, so the rate of entry of angular momentum to the turbine is:

$$\rho Q u_{ti} r_i$$

A similar argument gives us the rate at which angular momentum leaves the turbine:

$$\rho Q u_{to} r_o$$

The difference between these two is then the torque exerted on the runner by the fluid. So the torque is:

$$\Gamma = \rho Q(u_{ti} r_i - u_{to} r_o)$$

From this, we can find the power extracted from the water by the turbine:

$$P = \Gamma\omega = \rho Q(u_{ti} r_i - u_{to} r_o)\omega$$

The maximum torque and maximum power extraction happen when the second term in the brackets is zero, which means achieving $u_{to} = 0$. Looking back at the velocity diagram, this allows us to find the best angle for the trailing edge of the blade. This will be when:

$$\tan\beta_o = \frac{u_{ro}}{r_o\omega} \tag{2.9}$$

That completes the analysis. The design speed of the turbine is specified by the application. The power that can be extracted from a given reservoir is set by the design flow rate together with the available head. Regulating the flow is the way to control power output. The final SAQ will guide you through applying the formulae for one of the Dinorwig turbines.

SAQ 2.6 (Outcome 2.4)

The runner of each of the Francis turbines at Dinorwig has a diameter at the inlet of 3.80 m. The effective diameter of the blades at the outlet is 1.71 m. The effective height of the runner at the inlet (allowing for some obstruction by the guide vanes) is 0.230 m and 0.644 m at the outlet. The flow rate of water is 65 $m^3\ s^{-1}$. The runner revolves at 500 rpm when the guide vane angle is set at 13°.

(a) Find the optimum values for the two runner angles β_i and β_o.

(b) Determine the speed at which the turbine would operate most efficiently if the guide vane angle were changed to 17°.

5.5 Choosing a turbine

The choice between the impulse and reaction types of turbine requires a careful analysis of several factors. As the historical review demonstrated, one of the advantages of the Francis turbine is that it operates submerged and so can tolerate backing up of the outflow water. Another advantage is that all the turbine blades are used all the time, making a Francis turbine much more compact than the simple, single-jet Pelton wheel.

The output of either turbine depends on having a head to drive it, but the flow speed is determined by the swirling around the Francis turbine rather than the diameter of the nozzle of the impulse scheme. This means that reaction turbines can be efficiently operated from lower heads than impulse turbines, though they both do well on high heads.

As is often the case, the price to be paid for compact and efficient systems is in their capital cost. To calculate how effectively the efficiency savings recoup the investment requires a sound knowledge of patterns of usage, with reliable forecasts and no small measure of second sight – and all that is a business in its own right.

6 SUMMARY

In this part of Block 4, you have combined the historical study of the development of what is a hugely important machine with some analysis of how it works. This has enabled you to relate its efficiency to its physical form.

The aim of the historical section was to show a real example of how innovation and development in engineering takes many forms, and there is no best approach for all time and all circumstances. The way forward is determined by an interplay between the nature of the needs at any one time and the means at the engineer's disposal, which comprise both the theoretical background knowledge and the current manufacturing techniques and materials. In short, engineering is a cultural activity just as much as painting or music.

There are several basic mechanical ideas that you should take away from the mathematical analysis of water turbines A quick glance at an engineer's notebook (see Figure 2.37) and the following points summarize the essential knowledge.

- An impulse turbine operates with a jet hitting blades of the rotor. The rate of change of momentum of this jet equals the force of the jet on the blades.
- With no load, an impulse turbine spins up to an angular speed where the jet has no force on the blades (or just enough to keep the wheel turning). As load increases, the angular speed drops.
- A reaction turbine operates by a flow of water through the runner. The rate of change of angular momentum of the flow equals the torque exerted on the blades of the runner.
- A reaction turbine spins up to its design speed, where it is matched to its load by setting vane angles and flow rates.
- A reaction turbine can be used as a pump if operated in reverse.

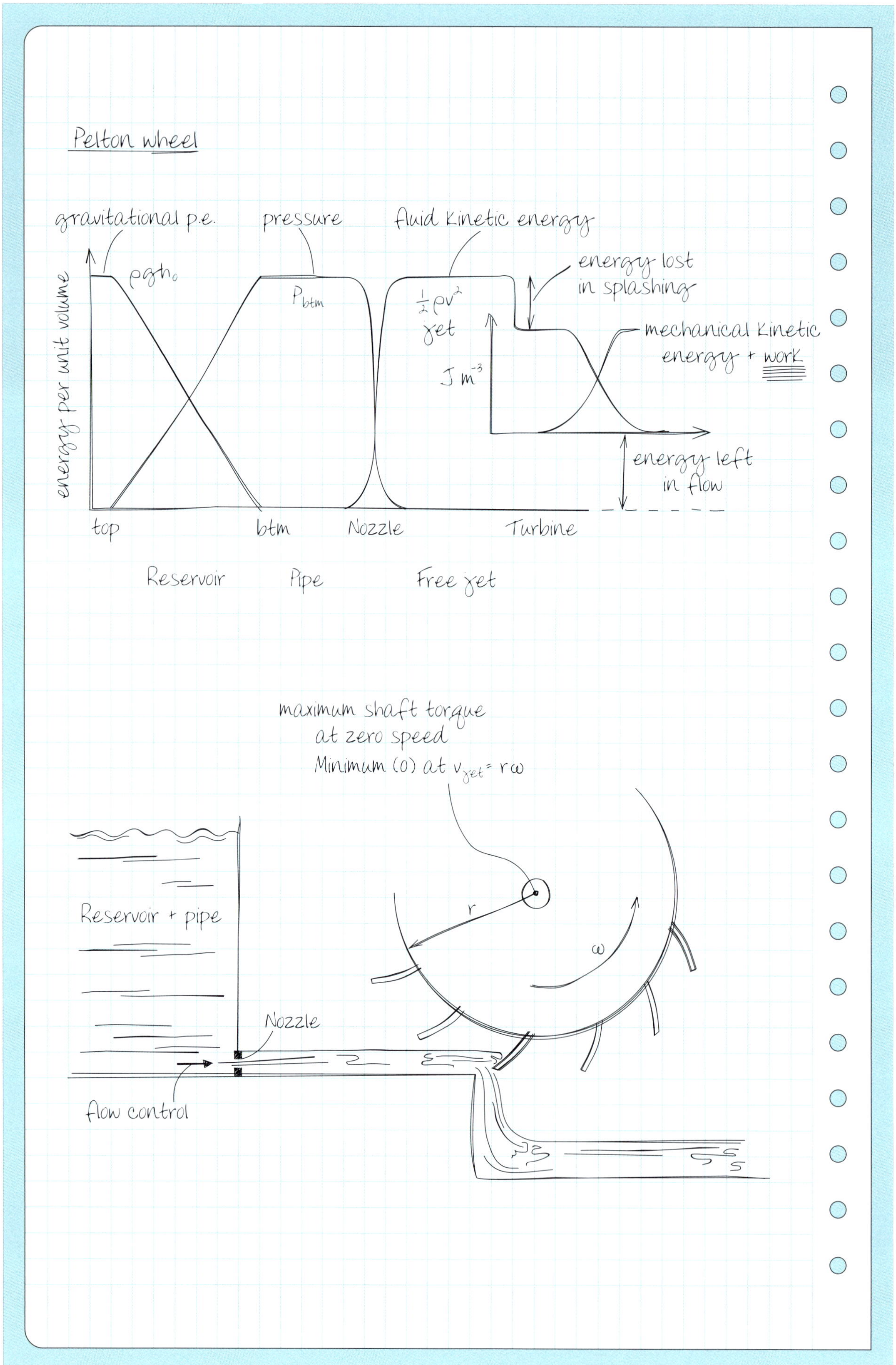

Figure 2.37 The Pelton wheel and the Francis turbine – pages from an engineer's notebook

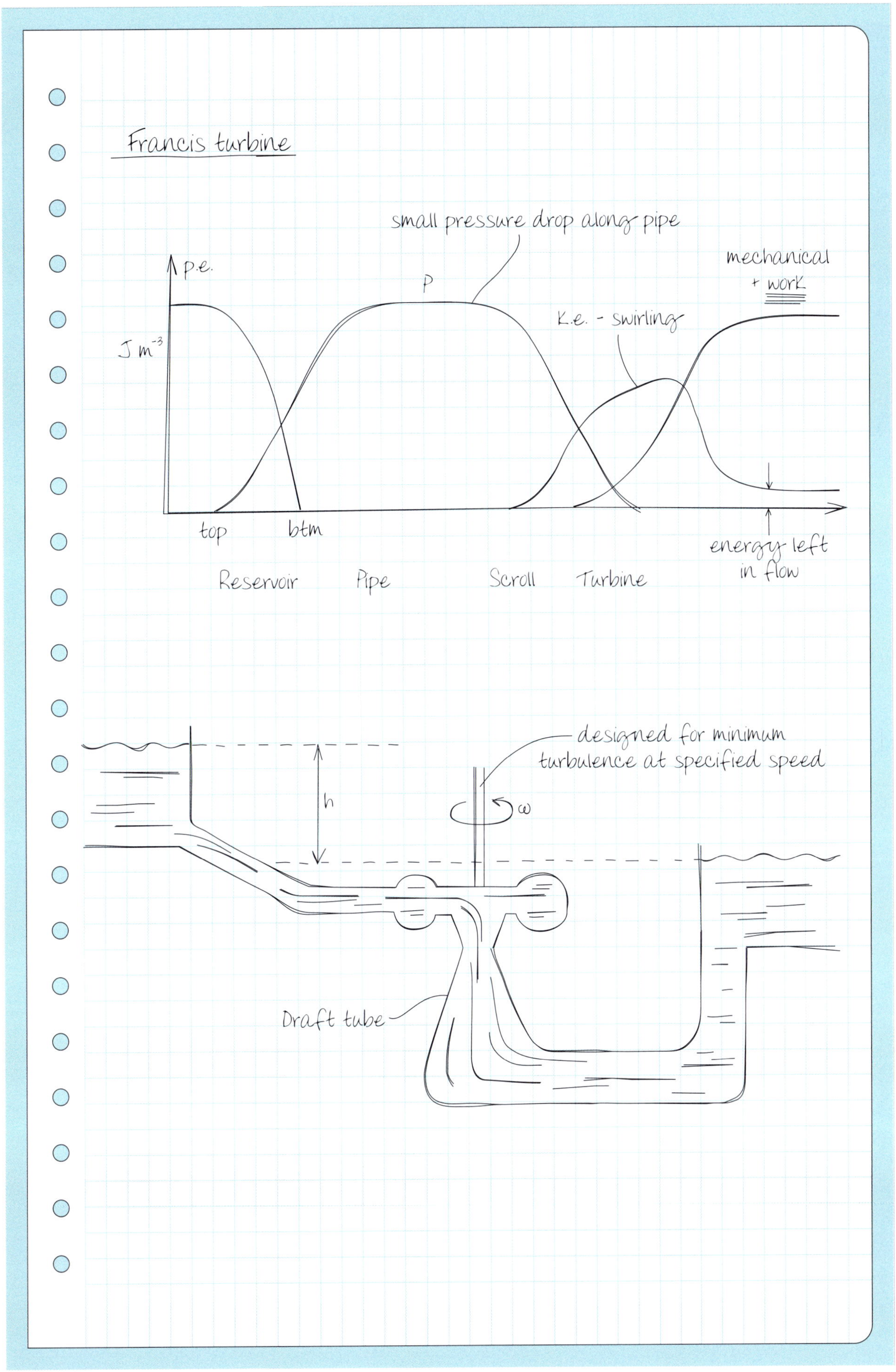

Figure 2.37 continued

7 LEARNING OUTCOMES

After studying Part 2 of Block 4 you should be able to do the following.

2.1 Explain the influence of cultural factors on the rate of technological development. (SAQ 2.2)

2.2 Give examples of how new materials and manufacturing techniques can affect technological developments. (SAQ 2.3)

2.3 Do calculations on the factors controlling the power output and efficiency of impulse turbines. (SAQs 2.4 and 2.5)

2.4 Do calculations on the factors controlling the power output and efficiency of reaction turbines. (SAQs 2.1 and 2.6)

ANSWERS TO EXERCISES

Exercise 2.1

1 *Independence:* Varying parameters one at a time will not work well if there is an interaction between the effects of two or more parameters. For example, the effect of oven type or temperature on the optimum baking time of a cake can be investigated by varying one or the other. However, the interaction between the amounts of water, fat and flour makes it more difficult to vary either one of these factors independently.

2 *Uniqueness:* There is no guarantee that the apparently optimum values found for a set of chosen parameters is not just a 'local maximum', with a better combination elsewhere in 'parameter space'. Tuning a car engine to run on pure alcohol will result in a combination of timings, fuel/air mixtures, compression ratios, etc. that can be shown to be optimal, but will certainly no longer be so if the fuel is changed to petrol.

3 *Missed parameters:* Without good theoretical knowledge of the underlying principles, there is no guarantee that *only* and *all* the important parameters have been selected for variation.

4 *Complexity:* The task quickly becomes large as the number of parameters to be varied increases, especially if some attempt is made to deal with interactions between parameters, for instance by covering all possible combinations of values for each.

Exercise 2.2

- Frictional losses slow the flow near surfaces, removing kinetic energy from it.
- Pouring from any height into the tail race means some potential energy is not being transferred into rotational energy.
- Whirls and eddies are a description of turbulence in which directed kinetic energy of flow is diverted into randomized kinetic energy that cannot be converted by the wheel to rotational energy.

Exercise 2.3

Return to the text and read on!

Exercise 2.4

The hydroelectric turbine was expected to generate power for a large number of users (possibly a whole town), not just a single enterprise (one manufacturing or processing business). The output power depends upon the head and the flow rate.

The high speed requirement follows from the way a generator works. Underlying the method is the principle of electromotive force being induced in proportion to the rate of change of a magnetic field. High turbine speeds ensure adequate rates of change of magnetic field.

ANSWERS TO SELF-ASSESSMENT QUESTIONS

SAQ 2.1

Torque

The torque, Γ, is given by:

$$\Gamma = \frac{P}{\omega}$$

So the missing entries in the table are:

(b) Torque = $(11.1 \times 10^3)/58.0 = 191$ N m

(c) Torque = $(24.9 \times 10^3)/62.8 = 397$ N m

Efficiency

The rate at which gravitational potential energy could be extracted from the water is given by:

$$Q \times g \times h = Q \times 9.81 \times 9.15 = 89.8Q \quad \text{(in watts)}$$

where Q is the flow rate in kilograms per second and h is the head in metres.

The efficiency of the turbine is therefore given by:

$$\frac{\text{power output}}{89.8Q} \times 100\%$$

So the missing entries in the table are:

(a) Efficiency = $100 \times (3.2 \times 10^3)/(89.8 \times 46.7) = 76.3\%$

(d) Efficiency = $100 \times (41.4 \times 10^3)/(89.8 \times 570.6) = 80.8\%$

SAQ 2.2

One cause of this is the evolution of needs: as they change, so does the motivation for improvement, as put so succinctly in the saying 'necessity is the mother of invention'. For hundreds of years, the needs were largely restricted to milling corn and raising water for irrigation. Efficiency was not the prime concern, as these processes could be carried out at a leisurely pace. As industrialization intensified, however, the rate at which power could be extracted from the water became increasingly important, as for instance in mineral extraction and preparation and in the metal smelting and processing industries. Once established, this need for power intensified, especially with the advent of electrical power distribution, which effectively linked the water wheel to many rather than just one user of the wheel's power. The steadily increasing 'pull' from the needs drives the general trend of acceleration in the rate of improvement of the water turbine.

The other contributor to this acceleration is the steadily increasing efficiency of the communication of ideas, through the learned societies, the patent system and increasingly efficient means of transport. An example of this is the way Fourneyron's, Boyden's, Howd's and Swain's patents were all used by others, who were able to make further improvements on these innovations.

The 'bumps' superimposed on this general trend can be explained in terms of the 'supply side' of the equation – in other words, the techniques employed by the innovators and improvers. The experimental methods pioneered by people such as Smeaton and de Parcieux enabled the accurate and quantitative comparison of wheels against one another, and their use of scale models also helped to increase the pace of development. A period of rapid improvement followed, which gradually slowed as these empirical methods faltered for lack of theoretical knowledge. Development was able to accelerate again with the injection of such knowledge through the work of Euler, Bernoulli and Borda. Another barrier to further development arose as the efficiency of the wheels came nearer to perfection; the assumptions and approximations made by the theoreticians to make the maths possible made their presence increasingly felt, such that further progress was better made by recourse once again to empirical methods – the period of 'cut-and-try' referred to in the text.

SAQ 2.3

Cast iron replaced wood, with the following advantages:

- better bearings;
- better dimensional stability, allowing manufacture to closer tolerances;
- stronger and longer-lasting wheels;
- complex three-dimensional shapes formed easily, allowing curved runners, scroll casings, etc. to be manufactured, thus making it easier to experiment with incremental design changes.

Wrought iron replaced cast iron, making possible the production of:

- lighter, tougher wheels;
- cheap sheet iron;
- strong, large-diameter water pipes for operation under high pressure.

The use of reinforced concrete permits the cheap manufacture of very large draft tubes for large hydroelectric turbines.

SAQ 2.4

Equivalent h_0 is $100 \times 10 = 1000$ m

$$v_{\text{jet}} = \sqrt{2gh_0}$$
$$= \sqrt{2 \times 9.81 \times 1000} = 140 \text{ m s}^{-1}$$

In more familiar terms, this is 504 km h^{-1}!

One reason the lances on these washers are so long is to make it difficult to accidentally turn the jet onto yourself.

SAQ 2.5

(a) The maximum torque is provided when the wheel is not rotating, i.e. $\omega = 0$ (because the velocity of the water relative to the wheel is at its highest). The expression for torque (Equation 2.5) then reduces to:

$$\Gamma = 2rv_{\text{jet}}^2 \rho A$$
$$= 2 \times 2 \text{ m} \times (30 \text{ m s}^{-1})^2 \times 1000 \text{ kg m}^{-3} \times 10^{-3} \text{ m}^2$$

So the maximum torque is 3600 N m.

(b) When there is no load, the wheel will spin up until the tangential speed of its blades matches the water-jet speed:

$$\omega = \frac{v_{\text{jet}}}{r} = \frac{30 \text{ m s}^{-1}}{2}$$

So the no-load angular speed of the wheel is 15 rad s^{-1}.

(c) Maximum power is delivered at half the no-load speed, and is given by Equation (2.6):

$$P = 2r\omega\,(v_{jet} - r\omega)\rho v_{jet}A$$

With $\omega = \omega_{max} = {}^{1}/_{2}\ v_{jet}/r$, this reduces to:

$$\begin{aligned} P &= \tfrac{1}{2}\rho {v_{jet}}^{3}A \\ &= \tfrac{1}{2}\times 1000\ \text{kg m}^{-3}\times(30\ \text{m s}^{-1})^{3}\times 10^{-3}\ \text{m}^{2} \end{aligned}$$

So the maximum power is 13 500 W or 13.5 kW.

(d) The relationship between torque and angular speed is a linear one. In (a) we found the torque for a stationary wheel to be 3600 N m, and we know that at the no-load angular speed of 15 rad s^{-1}, the torque provided by the wheel is zero. Therefore the angular speed at which it provides 2200 N m is:

$$15\times\left(1-\frac{2200}{3600}\right) = 5.83\ \text{rad s}^{-1}$$

With $r = 2$ m, $r\omega = 11.66$ m s^{-1}.

From Equation (2.6), the power output is:

$$2\times 11.66\times(30-11.66)\times 1000\times 30\times 10^{-3} = 12\,831\ \text{W}$$

The theoretical maximum power output we found in (c) represents 100% efficiency, therefore the efficiency is 12 831/13 500 × 100 = 95%

SAQ 2.6

There is enough information here to allow us to calculate the magnitude of the radial component of the absolute water flow velocity at inlet, u_{ri}. It is given by:

$$\begin{aligned} u_{ri} &= \frac{Q}{\pi d_i Z_i} \\ &= \frac{65}{3.14\times 3.80\times 0.230} = 23.7\ \text{m s}^{-1} \end{aligned}$$

Similarly, at outlet:

$$\begin{aligned} u_{ro} &= \frac{Q}{\pi d_o Z_o} \\ &= \frac{65}{3.14\times 1.71\times 0.644} = 18.8\ \text{m s}^{-1} \end{aligned}$$

Since we know the guide vane angle, α, we can get the tangential component of the water's velocity at inlet using Equation (2.7):

$$\begin{aligned} u_{ti} &= \frac{u_{ri}}{\tan\alpha} \\ &= \frac{23.7}{0.23} = 103.0\ \text{m s}^{-1} \end{aligned}$$

The angular velocity of the runner blade is given by:

$$\omega = \frac{500\times 2\times\pi}{60} = 52.4\ \text{rad s}^{-1}$$

We can also find the tangential velocity of the runner blade at inlet, as it is the angular velocity times the radius:

$$r_i\omega = 1.9 \times 52.4 = 99.6 \text{ m s}^{-1}$$

(a) The optimum blade angle at inlet is given by Equation (2.8):

$$\tan\beta_i = \frac{u_{ri}}{(u_{ti} - r_i\omega)}$$

$$= \frac{23.7}{103.0 - 99.6} = 6.97$$

So the optimum blade angle at inlet, β_i, is 82°.

At the outlet, the optimum angle is such that $u_{to} = 0$, so Equation (2.9) gives:

$$\tan\beta_o = \frac{u_{ro}}{r_o\omega}$$

$$= \frac{18.8}{0.855 \times 52.4} = 0.42$$

So the optimum blade angle at outlet, β_o, is 23°.

(b) If the guide vane angle is changed to 17°, this will change the value of u_{ti} to $u_{ri}/\tan 17° = 77.5 \text{ m s}^{-1}$.

For optimum efficiency, the runner will adjust its speed to keep the blades tangential to the water flow at the inlet. Therefore, we can find the new value of ω by substituting the new value of u_{ti} into the expression for β_i. Using Equation (2.8):

$$\tan\beta_i = \frac{u_{ri}}{(u_{ti} - r_i\omega)}$$

gives:

$$6.97 = \frac{23.7}{77.5 - 1.90\omega}$$

We can rearrange this to find the value of ω that keeps the blade at inlet parallel to the water flow:

$$\omega = \frac{77.5 - \dfrac{23.7}{6.97}}{1.90} = 39 \text{ rad s}^{-1}$$

The turbine would therefore operate most efficiently at 39 rad s^{-1} (372 rpm). Notice now that the angle of the blade at outlet is no longer ideal, as a slight tangential component has appeared in the water flow. This is why the efficiency of the turbine falls off as the guide vane angle is changed from the design setting.

REFERENCES

Freeman, J. P. (1923) 'The fundamental problems of hydroelectric development', American Society of Mechanical Engineers, New York.

Hunter, L.C. (1979) *A History of Industrial Power in the United States, 1780–1930*, vol. 1, 'Water power in the century of the steam engine', University Press of Virginia, Charlottesville.

Keator, F. W. (1939) 'Benoît Fourneyron (1802–67)', *Mechanical Engineering*, vol. 61.

Layton, E. T. (1979) 'Scientific technology, 1845–1900: the hydraulic turbine and the origins of American industrial research', *Technology and Culture*, vol. 20.

Reynolds, T. (1983) *Stronger than a Hundred Men: A History of the Vertical Water Wheel*, Johns Hopkins University Press, Baltimore and London.

Safford, A. and Hamilton, E. (1922) 'The American mixed-flow turbine and its setting', *Transactions of the American Society of Civil Engineers*, vol. LXXXV.

Smith, N. (1976) *Man and Water: A History of Hydro-technology*, Peter Davis, London.

ACKNOWLEDGEMENTS

Grateful acknowledgement is made to the following sources for permission to reproduce material within Part 2 of this block.

Figures

Figure 2.1: Allis-Chalmers; *Figure 2.2:* Courtesy of Gilbert Gilkes & Gordon Ltd; *Figure 2.9:* The Ambrosiana Library, Milan; *Figure 2.13:* Wilson, D.G. 'Turbo machinery from paddle wheels to jets', *Mechanical Engineering,* Vol. 104, No. 10, October 1939; *Figure 2.14:* Keator, F.W. 'Benoît Fourneyron 1802–1867', *Mechanical Engineering,* Vol. 61, No. 4, April 1939; *Figure 2.15:* Courtesy of the Lowell Historical Society; *Figure 2.16:* With permission from The American Textile History Museum, MA; *Figure 2.20:* Reynolds, T. (1983) *Stronger than a Hundred Men: A history of vertical water shed*, Johns Hopkins University Press; *Figure 2.22:* The Science Museum/Science & Society Picture Library; *Figure 2.23:* The Centre for Alternative Technology; *Figure 2.25:* Courtesy of James Leffel & Company, Springfield, Ohio.

Every effort has been made to contact copyright owners. If any have been inadvertently overlooked, the publishers will be happy to make the necessary arrangements at the first opportunity.

Course team acknowledgements

Part 2 of this block was prepared for the course team by Jan Kowal, with contributions from Chris Earl. It is largely based on T362 *Design and Innovation* Block 1 Section 2 by Mike Flood.

Part 3
Bearing the load

CONTENTS

1 INTRODUCTION

1.1 Aims

The overall aims of Part 3 of this block are:

- To outline the fundamental concepts and principles of friction, wear and lubrication in an engineering and industrial setting.
- To illustrate how engineers deal with friction, wear and lubrication problems using the context of hydroelectric power generation plant.

1.2 The importance of bearings

Dinorwig Power Station is the largest pumped-storage power station in Europe, located inside a mountain adjacent to the Snowdonia National Park in north Wales and operated by Edison Mission Energy. At times of low demand for electricity on the national grid, water is pumped to a high-level reservoir by reversible pump-turbines, and, at times of peak demand, electricity is generated by the water passing through the pump-turbines as it falls under gravity.

In action, each of the six turbine and generator sets at Dinorwig contains some 500 tonnes of rapidly spinning components.

How do you hold this mass in place and support it whilst still leaving it free to rotate?

The answer is 'You use bearings'.

Bearings have been part of our society for a long time. Arguably, the most important step in the invention of the wheel was not the wheel itself but the realization that something was needed to hold the wheel in place while it rotated, i.e. the bearing for its shaft or axle.

Exercise 3.1

Think about the bearing of a wheel on a wooden cart a thousand years ago. Say it has a fixed axle, at either end of which a wheel is mounted with a bearing. Other than securing the wheel to the axle, what other functions does the bearing perform? Name the physical phenomenon that is necessary for the wheel to rotate on the ground and whose absence is necessary for the shaft to rotate in the bearing.

So the object of bearings is to provide support at the same time as minimizing losses due to friction. The main energy-loss process is wear between the two sliding surfaces, which generates heat. This heat is usually dissipated to the surroundings and so is mostly irrecoverable. An important way to minimize friction and its associated wear is through lubrication (this was recognized as far back as in ancient Egypt). Together, the three areas of friction, wear and lubrication represent the principal planks of the subject called *tribology*. You'll be looking at aspects of tribology after the next section which examines the Dinorwig turbine plant design in some more detail. Sections 4 to 6 look at different aspects of bearings and their lubrication, focusing on lubricants, viscosity and types of bearing. Sections 7 and 8 then return to Dinorwig to consider issues relating to kinetic energy and energy losses. Finally, Section 9 discusses the phenomenon known as hydrodynamic lubrication.

Remember that 'work done' and 'energy' have the same units, newton metres or N m, which is the same as joules (1 N m = 1 J), so work done in overcoming friction represents energy lost or 'consumed'.

2 THE DINORWIG DESIGN

The heart of each power-generating set at Dinorwig is the massive rotating shaft assembly, supported via bearings by the strong and stiff support structures and foundations of the plant.

Figure 3.1 shows the cross-section of one of the six vertical-shaft power-generating and pumping sets. The inlet water arrives at the Francis turbine mounted at the lower end and imparts kinetic energy to the turbine runner. This is connected via an intermediate shaft to the rotor of the generator at the top of the assembly, which generates the electrical power. When in pumping mode, the generator becomes a motor which uses electrical energy (off-peak) to drive the turbine runner as a pump to impart kinetic energy to the water and send it back up to the upper reservoir.

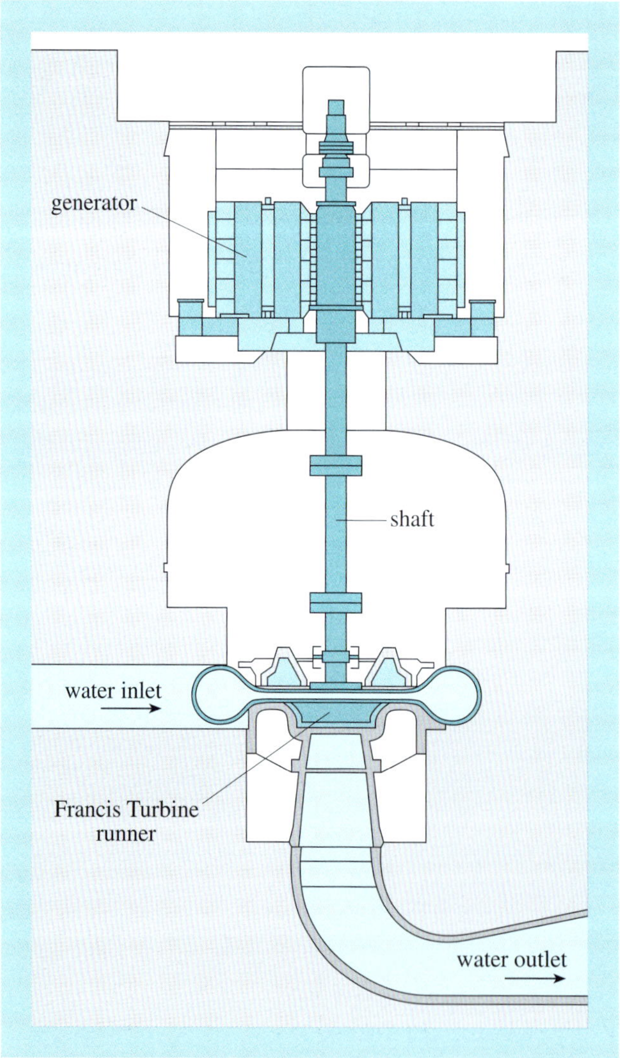

Figure 3.1 Cross-section of one of the six vertical-shaft generating and pumping sets at Dinorwig

Figure 3.2 shows the turbine/pump assembly, and Figure 3.3 the generator/motor assembly, both figures being part-sectioned. It all looks pretty complicated. Let's therefore make a start by simplifying things and looking at the main principles. Look at the rotating parts, starting at the bottom. There is the turbine runner, or rotor itself, connected to the turbine shaft. The turbine shaft is connected to an intermediate shaft and this is then connected to the generator rotor shaft. (Sounds like the old song!) The generator rotor shaft includes the rotor core assembly and an extension piece at the top containing the brush-gear. There is also a large collar forming part of a so-called thrust bearing which supports the weight of the whole rotating assembly. Thrust bearings carry or support axial loads, i.e. loads in a direction along the shaft axis. In this case, because of the vertical orientation of the plant, the weight is an axial load, along the main longitudinal axis.

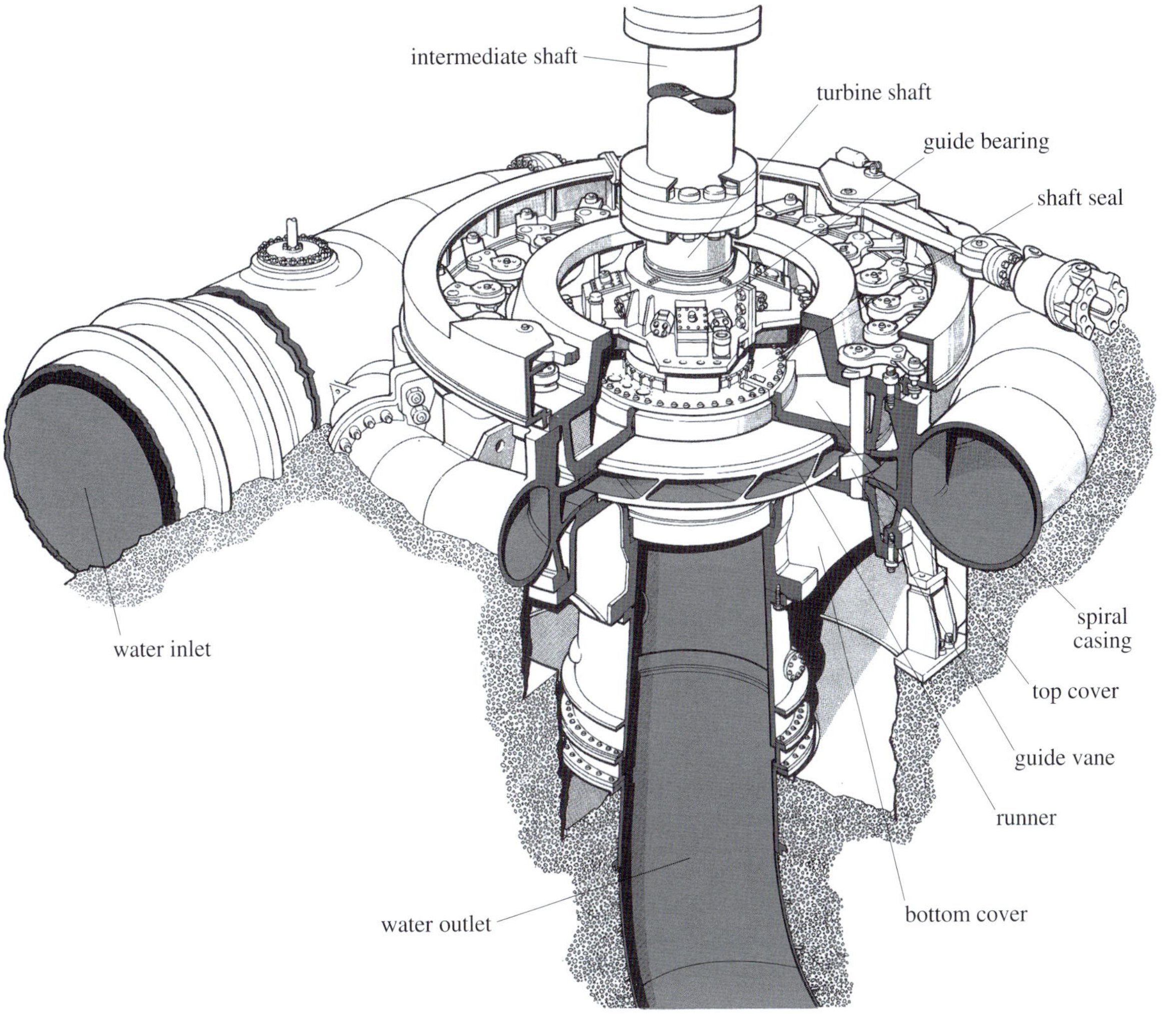

Figure 3.2 Turbine/pump assembly, Dinorwig

The rotating assembly has a mass of nearly 510 tonnes (1 tonne = 1000 kg) and hangs from the topmost thrust bearing. This bearing then is carrying a load of over 5 MN. Figure 3.4 shows a simplified sketch of the shaft assembly and its support bearings. As stated above, the thrust bearing at the top carries the vertical load (the weight) and can be thought of as a fixed annular support ring, through which the shaft passes, with the collar attached to the shaft transmitting the load to it. There is a thin film of oil between the support ring and collar, supplied by a circuit incorporating pump, filters, cooler, etc. The oil flows through the small gap and escapes at the free edges to be collected and recirculated.

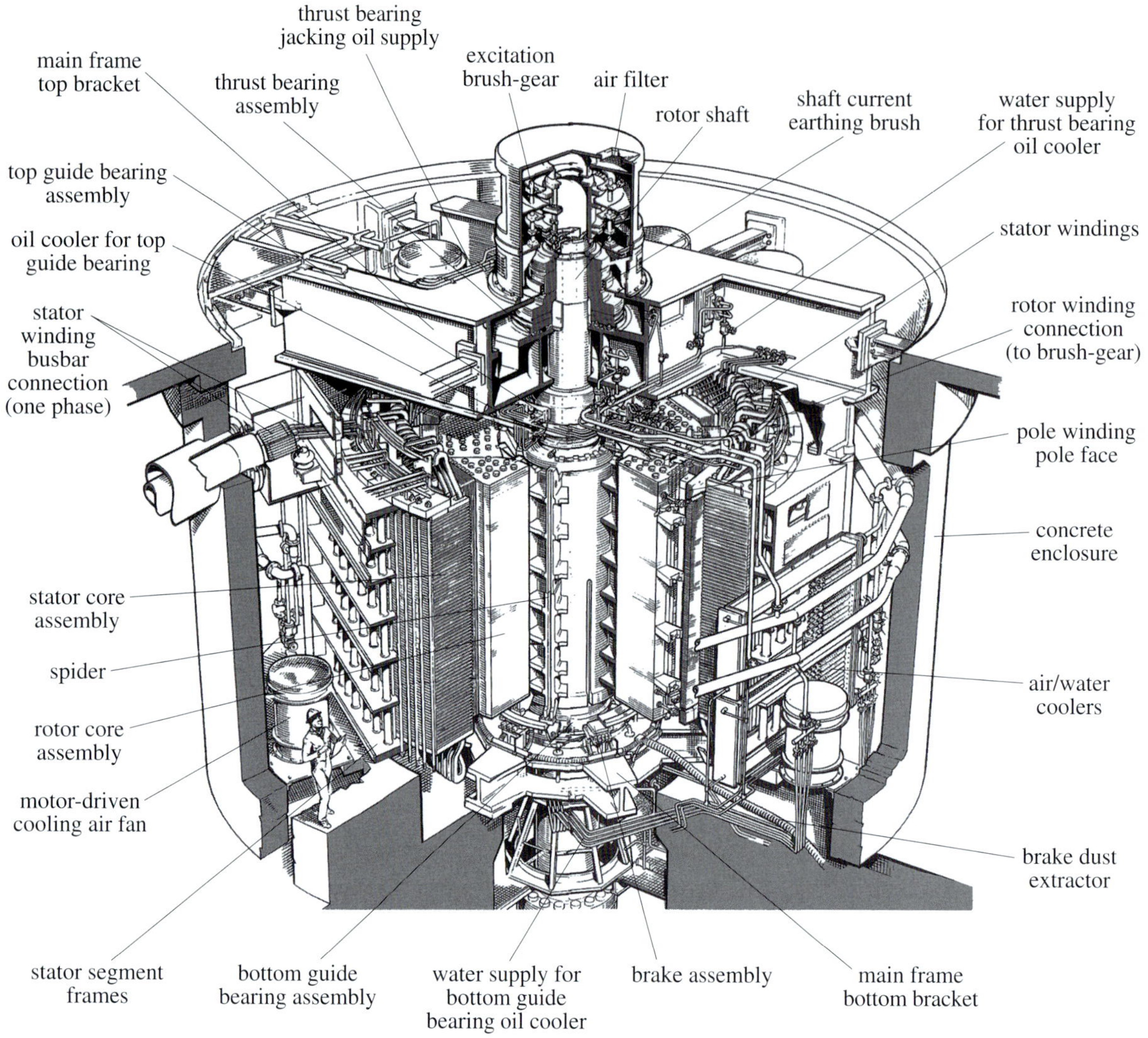

Figure 3.3 Generator/motor assembly, Dinorwig

Additionally, there are three guide bearings locating the shaft sideways and carrying any radial loads, for example the reactions from turbine and pump forces which act horizontally at the bottom. These guide bearings are also shown in Figure 3.4 and are journal-type bearings. For the moment these can each be thought of as a fixed collar surrounding the shaft (i.e. the journal) with an oil film in between. Again, the oil film is maintained by a supply circuit incorporating pump, coolers, filters, etc., and the oil which is allowed to escape from the free edges is collected and recirculated in the system. Additionally, the Dinorwig journal bearings are partly immersed in a bath of oil to ensure that some supply is always maintained. (If, like me, you're curious about why they're called journals, both the Oxford English Dictionary and Chambers Dictionary say the origin of the term is unknown. It is tempting to think that it was originally applied to bearings whose regular lubrication was so important that this had to be recorded in a log or journal.)

At Dinorwig, then, bearings provide the vital supporting role for the rotating mass that is at the heart of the power plant. The efficiencies of their design and operation are also crucial in maintaining reliability and reducing energy wasted, and hence costs, in running the plant. To appreciate their working requires you to explore some of the fundamentals that underpin our understanding of bearings. Through such an understanding come ways and ideas to improve performance and efficiency of operation, particularly in terms of energy losses. You'll see how this has important implications for the Dinorwig plant and its ability to cope with operating conditions that differ from those for which it was originally designed.

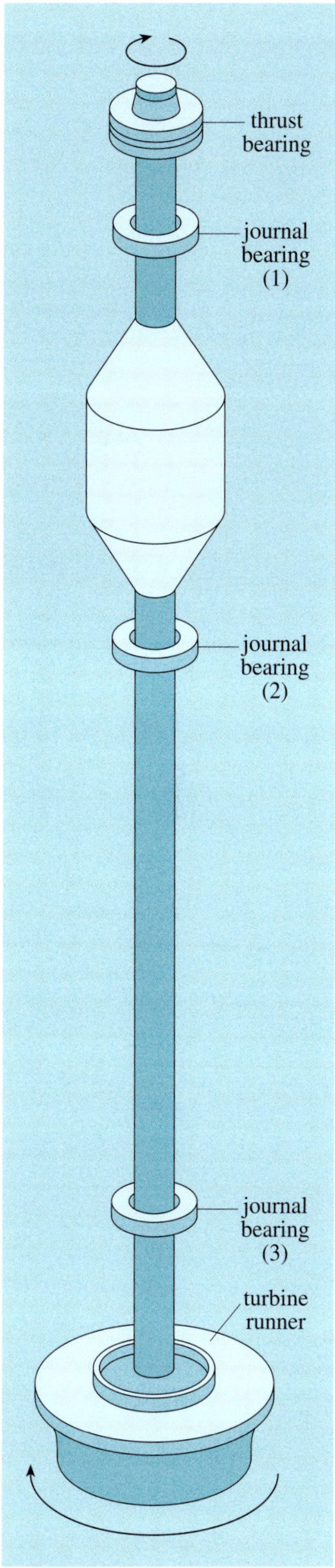

Figure 3.4 Vertical shaft and bearings, Dinorwig

3 TRIBOLOGY BASICS

Tribology (from the Greek 'tribos', meaning 'rubbing') has been defined as 'the science and practice of interacting surfaces in relative motion and of the practices related thereto'. As I said earlier, friction, wear and lubrication are the three core aspects of the subject. It received a lot of attention (though not under the name 'tribology') from quite early times and a number of distinguished workers have conducted experiments on friction and lubrication. These included Leonardo da Vinci in 1508, Newton in 1687, Euler in 1748 and Coulomb in 1809, all of whom formulated ideas on the subject that are still relevant today.

It has been estimated that between a third and a half of the total energy produced in the world is consumed in overcoming friction. This may seem a startling claim, but think of all the cars, lorries, buses, trains, aeroplanes, ships, production line machinery, machine tools, pumps, computer disk drives and the like in the world. They all expend a significant amount of power in overcoming friction. The car, for example, delivers useful work only after the friction of the engine crankshaft, valves, piston rings, cams, etc. have deducted their toll. It doesn't stop there, either. This useful work is then largely consumed in transmission gears and shaft bearing friction, rolling resistance of the tyres and even aerodynamic resistance (which is the equivalent of friction due to the air). Even very small reductions in the amount of friction (e.g. through more effective lubrication and improvements in bearing design) will have significant knock-on effects in terms of energy saving.

There have been a number of initiatives devoted to this theme. In 1966, a UK government-sponsored report concluded that the potential savings to British industry resulting from lubrication education and research were about (then) £515 million per annum. (The report also introduced the concept of 'Tribology' as a unifying subject title.) In 1992, the Institution of Mechanical Engineers launched a Tribology Action Campaign highlighting potential savings of £1.5 billion per annum in the UK. It was estimated that individual companies could save up to 10% each year on their energy bills by relatively simple reviews of basic lubrication practices and methods, i.e. not requiring major capital investment. Added benefits would accrue from reductions in downtime, labour costs and replacement plant.

3.1 Friction

Friction, then, is both a blessing and a curse. Without it, we couldn't walk, our cars wouldn't brake or go round corners, and boy scouts couldn't light fires. Many machines depend on it, but many more rely on minimizing it through lubrication.

So what is it?

Friction occurs whenever two solid surfaces in contact move, or attempt to move, relative to one another. It arises because such surfaces are never absolutely smooth. If you could examine the contact region between two surfaces at high magnification, you would find that, however smooth and flat they might appear to the eye, they were actually microscopically rough (see Figure 3.5). Even highly polished surfaces have hills and valleys several hundred atoms high. This means that the surfaces are in contact only at a series of points, called asperities, so that the actual area of contact is much smaller than the area of surface. The local stresses at asperities are thus very much higher than the overall stress (force/area) holding the surfaces together, high enough to cause plastic deformation, fracture and even welding together at the contact points. Friction is the force needed to overcome such effects. This is explored further in **▼A few areas of friction▲**.

You can see that this picture of friction also provides a mechanism for wear, the removal of material when surfaces rub together, which is the topic of the next section. Friction is associated with one thing sliding over another, and that is not something restricted to solids. In fluids, one region of fluid may well slide into and through another. The aerodynamic drag forces encountered in Block 3 Part 4 are effectively friction at work in air, and later in this part you will be meeting friction within lubricating fluids.

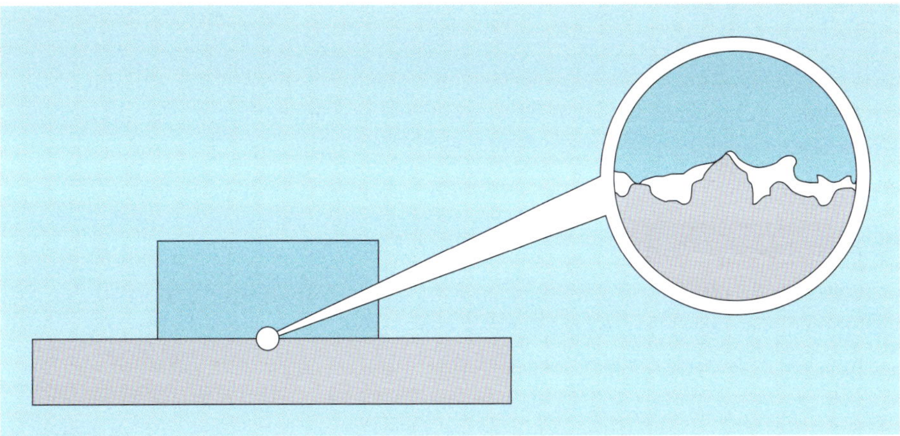

Figure 3.5 Schematic close-up of contact between surfaces

▼A few areas of friction▲

To investigate friction you could use an arrangement like that shown in Figure 3.6(a), in which you pull horizontally on a spring balance (a simple load cell) attached to a block of material resting on a flat plate of the same or different material. As you gradually increase the magnitude of the force F, you find that the block remains stationary until a particular value of F, when it starts to move. What does this mean?

It means that, at rest, F must be balanced by a force, shown as F_f in Figure 3.6(a), to maintain equilibrium. This is the frictional force. It increases with, and in reaction to, the tractive force F up to the point where movement starts. At this stage the frictional force has reached its maximum value, known as the *limiting frictional force* (or *limiting friction*), F_{lim}, Figure 3.6(b).

What happens once the block has started to move?

What you'd find is that the magnitude of F_f changes from F_{lim} to a lower steady value, F_d, known as the *dynamic frictional force* or *kinetic frictional force*.

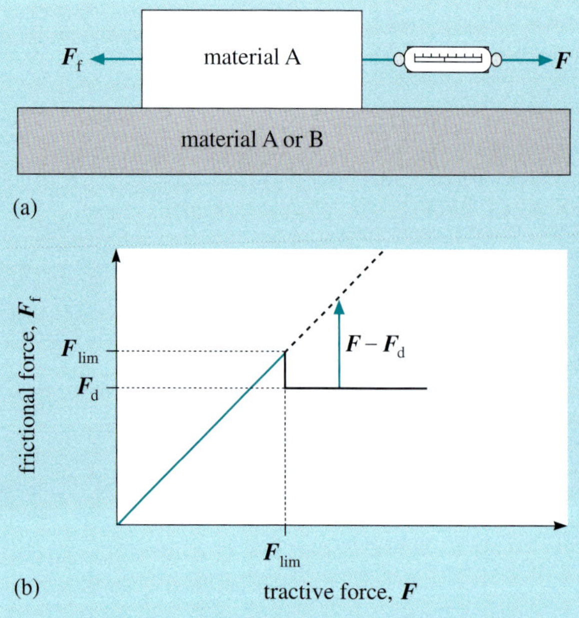

Figure 3.6 Investigating friction: (a) simple experimental set-up and (b) frictional force

If you relieve some of the contact area, but effectively retain the same mass (see Figure 3.7a), and repeat the experiment, you'll find that, for many materials, the frictional force and hence the values of F_{lim} and F_d are unchanged. For these, the frictional force is independent of area of contact. If, on the other hand, you increase the weight of the block, as sketched in Figure 3.7(b), you'll find that the frictional force increases in direct proportion to the weight of the block.

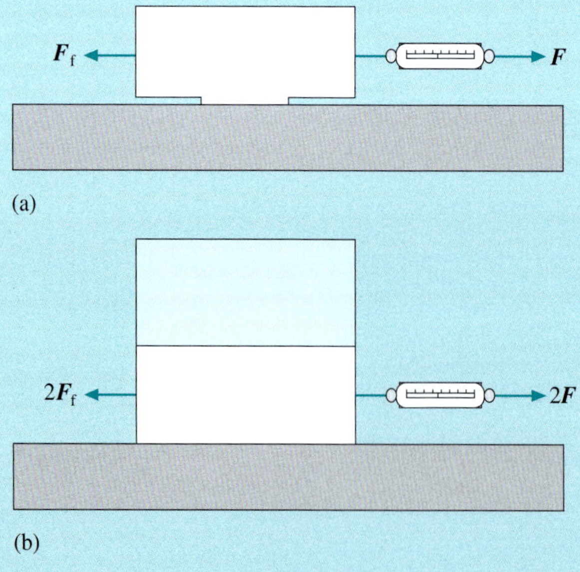

Figure 3.7 Limiting friction: (a) effect of area and (b) effect of adding weight

What's important in determining frictional forces is the extent to which surfaces are being pushed together. By Newton's Third Law, the weight of the block is balanced by an upward reaction force, R, of the surface pushing on the block and equal in magnitude to the weight (see Figure 3.8a). This reaction force acts perpendicularly to the surface and hence is known as the *normal reaction*. It follows that on non-horizontal surfaces the reaction normal to the surface that is effectively acting to push the surfaces together is equal to the component of the weight perpendicular to the surface (see Figure 3.8b). You can then generally write $F_{\text{lim}} \propto R$, or:

$$F_{\text{lim}} = \mu_s R \qquad (3.1)$$

where μ_s (Greek 'mu') is the *coefficient of static friction*. The value of μ_s varies with the materials in contact and is usually in the range 0.2 to 1.0. It is also sensitive to surface contamination. Some typical data is listed in Table 3.1. You can see the effect on μ_s of lubricating the surfaces with suitable liquids.

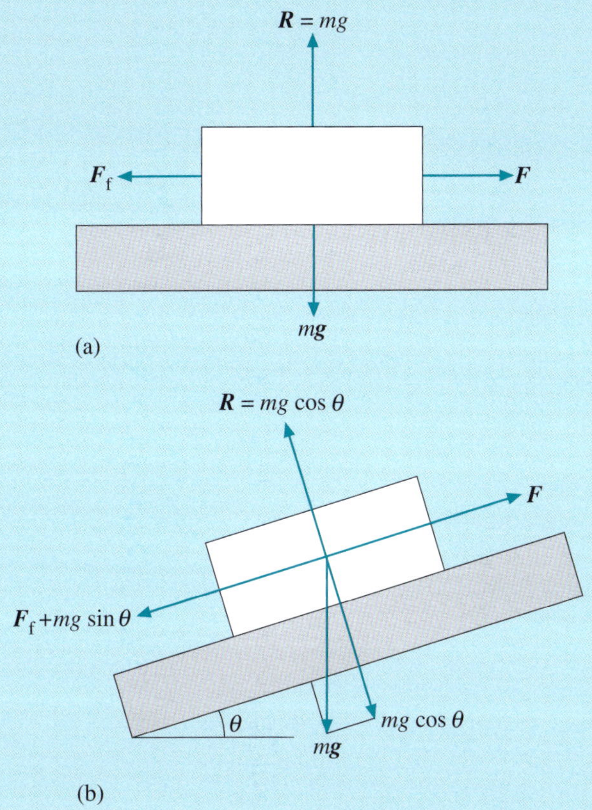

Figure 3.8 Forces acting on a block subjected to a force: (a) along a horizontal plane, and (b) on an inclined plane. The limiting value of F_f is $\mu_s mg \cos\theta$

As with static friction, the dynamic frictional force is independent of area but is proportional to weight (i.e. R). With many materials (though not polymers), it is also independent of sliding speed. In the same way as for static friction, a *coefficient of dynamic friction*, μ_d, can be defined as:

$$F_d = \mu_d R \qquad (3.2)$$

Table 3.1 Some typical values of the coefficient of static friction, μ_s

Surfaces in contact	Contact conditions	μ_s
Most metals on self	unlubricated, in air	1.0
	lubricated, mineral oil	0.2–0.4
Hard steel on self	unlubricated	0.6
	lubricated, mineral oil	0.14–0.2
	lubricated, molybdenum disulphide	0.1
Glass on self	clean, in air	1.0
	lubricated, liquid hydrocarbons	0.3–0.6
Ice on self	below –50 °C	0.5
	between 0 °C and –20 °C	0.05–0.1
Typical wood on self	clean, dry	0.25–0.5
	clean, wet	0.2
Perspex®, polyethylene* on selves	clean	0.8
Perspex®, polyethylene on steel	clean	0.3–0.5
Nylon on self*	dry	0.5
PTFE* on steel or self	dry	0.04–0.1
Rubber tyre on road	dry	1.0
	wet	0.6
Ski on snow	at 0 °C	0.05
Brake lining on brake drum	clean	0.4
	wet	0.2
	greasy	0.1

* Note that with polymers, μ_s decreases with increasing load and μ_d (coefficient of dynamic friction) increases with increasing speed

SAQ 3.1 (Outcome 3.1)

(a) What horizontal force is required to just move a hard steel block of mass 5.0 kg on an unlubricated, horizontal, hard steel surface?

(b) At what angle of slope would skiers on snow just begin to move downhill under their own weight when the temperature is 0 °C?

(c) A car of mass 800 kg and moving at 50 km h^{-1} is required to stop in 7 s. If the only retarding force is that provided by the friction between the drums and linings of the four brakes, and each brake contributes equally to this force, estimate the force that must be applied between each drum and its lining to stop the car if the coefficient of dynamic friction is 0.3.

3.2 Wear

One of the most widespread problems in engineering is how to allow two surfaces to rub together without causing damage to either one. This ranges from mundane applications like rubber tyres on a tarmac road through to ceramic/polymer interfaces in prostheses.

Because wear is something that will eventually lead to failure of a component – either because it will become loose in its fitting owing to loss of material or because it may lose its ability to bear a design load – it is a mechanism of failure that needs to be understood by a design engineer. In the sense that undesirable wear represents failure of the material, the topic has obvious affinities to Block 5. However, its close links with friction place it here. Although friction and wear go hand in hand, they are by no means the same thing. You'll see the difference between the two if you look at the development of **▼Prosthetic hip joints▲**.

▼Prosthetic hip joints▲

Many of us know someone who's undergone surgery for a hip replacement. Over 40 000 hip replacements are performed in Britain each year, so it is becoming a fairly routine surgical procedure.

The natural hip is a ball-and-socket joint, the ball being the head of the femur – the thigh bone – and the socket being the cup in the pelvis called the acetabulum. Arthritis causes the lubricating surfaces, made of cartilage, to degrade, and eventually bone rubs upon bone. This is the cause of stiffness and pain. In the replacement operation, all the damaged bone and cartilage is removed. The entire natural head of the femur is cut off and replaced by a metal head with a stem that fits down the central cavity of the femur. Stable fixation is a prime consideration and the prosthesis is often cemented into place using an acrylic cement packed into the femoral cavity before the replacement is inserted and then allowed to cure *in situ*. The socket is relined with a cup – usually a polymer – which is cemented into place (see Figure 3.9).

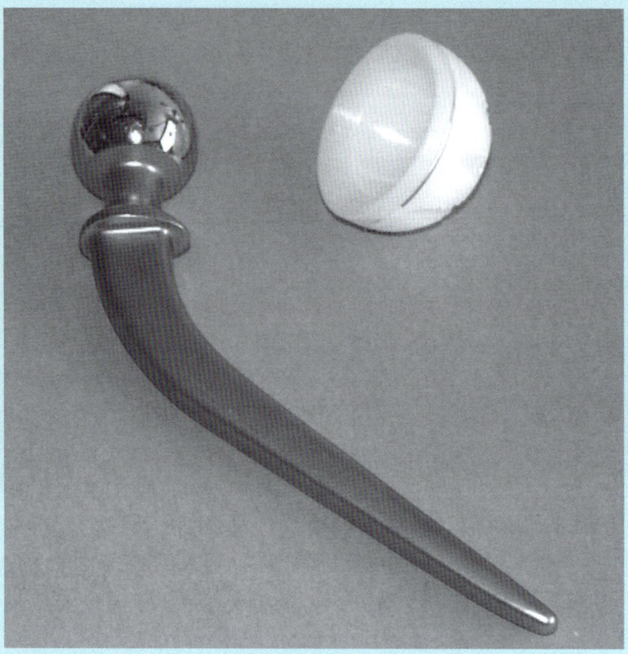

Figure 3.9 The components of a prosthetic hip joint

This produces an artificial joint of a metal bearing on a polymer. This is the right 'engineering' combination for a low coefficient of friction, but one that is nonetheless still over ten times higher than that of the original cartilage-on-cartilage bearing. Don't underestimate Mother Nature's superiority in materials design!

The key functional requirements of a replacement hip are disparate and demanding. The ideal hip replacement would give pain-free normal mobility for the lifetime of the patient.

The joint may undergo fairly severe cyclic loading: for example, it may have to withstand occasional transient loads of up to 50 times body weight when landing from a jump and cope with normal walking loads of 2.5 times body weight applied 10^6 times a year. In addition, there must be no possibility of catastrophic failure, such as that caused by rapid crack propagation from a defect.

The internal biological environment is chemically aggressive and any material chosen for long-term implantation must be able to resist these conditions without any reduction in mechanical properties or the release of undesirable degradation products into the body.

The coefficient of friction between the head and the cup affects the ease of movement of the joint and it should therefore be as small as possible. Friction also affects the rotational torque that is generated between the prosthesis and the bone, and thus the stability of the implant in the bone.

Exercise 3.2

Suggest from your general knowledge a 'low-friction' material that might be used in the cup of the hip joint.

PTFE (Teflon®) was a favoured choice in the early days of hip replacements. However, nothing could illustrate more starkly the difference between friction and wear. Have a look at Figure 3.10.

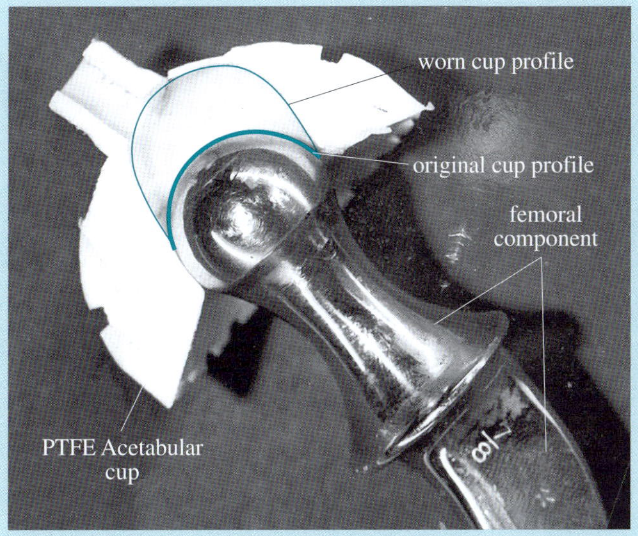

Figure 3.10 A worn Teflon® cup

Teflon® wears very rapidly, despite having a low coefficient of friction with metals. It is also not very passive within the body, causing inflammation, particularly from the tiny particles of debris worn away from the joint.

Wear of the prosthesis is therefore a very important consideration. The conventional modern hip replacement comprises a cup made from a polyethylene with an ultrahigh molecular mass and a titanium alloy (Ti–6% A1–4%V) head. This is an alloy which has a very high resistance to corrosion.

Such a metal-on-plastic prosthesis may wear very little over twenty years in an older patient, but the wear rate can be much greater in a more active young person and such a patient requires their hip to last maybe 50 years.

The new generation of hips for long-term use in younger patients has ceramic heads made of high-density alumina, which are attached to a titanium alloy stem. When work on this hip started in Germany, alumina was already being used in dental prostheses. It was therefore known to be safe for implantation. The design standard ISO 6474:1981 requires the material to be capable of enduring a lifetime of thirty years, loaded one to two million times per year. These stringent requirements are met by the careful design and control of the processing conditions, which affect the microstructure and hence the properties of the final component. (Microstructure will be discussed in Block 5 Part 2.)

Wear can be defined as the progressive loss of material from surfaces that have been rubbed against one another. In many components such as bearings, or any application where there are moving surfaces in contact with each other, wear can occur. The cost of wear, in terms of replacement of parts and associated loss of production time, was estimated to contribute to 1% of the US GDP in 1998.

You saw earlier that all surfaces, even those which have been finely polished, are rough at the microscopic scale. For example, Figure 3.11 shows an image of a 'smooth' surface taken with a stylus profilometer – rather like dragging a record player stylus across the surface of a material and measuring the vertical deflections. The magnification of this image is larger in the vertical scale than the horizontal. The surface has considerable variation, as can be seen from the peaks and troughs. The peaks are the asperities identified earlier. They are important in wear since sharp asperities on a hard material may plough into a softer material and cause gouging.

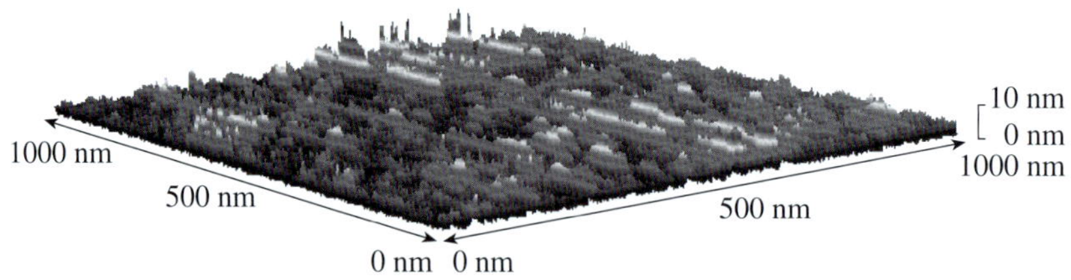

Figure 3.11 Profilometer trace of a 'super-smooth' silicon wafer (the magnification of this image is larger in the vertical scale than the horizontal). The vertical distance between the peaks and the troughs is about 10 nanometres

Also, when two surfaces are in contact, the true contact area can be very small, as the surfaces 'balance' on the points that are highest. Placing two surfaces in contact is rather like turning Yorkshire upside down and placing it on the Cotswolds. True contact will occur only where the tips of the 'hills' on one surface touch those on the other surface. These 'hills' will deform or fracture until the areas of contact are large enough to support the applied load. The details depend on the shapes of the asperities and on the elastic and plastic properties of the materials. This means that the wear – and the friction – can be controlled by how the asperities bear the load and deform.

Wear can be classified according to which of four types of mechanism is operating: abrasion, adhesion, fatigue or chemical action. These are explored in **▼Mechanisms of wear▲**.

▼Mechanisms of wear▲

Abrasive wear

Abrasion is one of the most common forms of wear. As well as being a potential problem, it is often exploited intentionally. Examples of this are using glasspaper (commonly known as sandpaper) to prepare wood surfaces, or silicon carbide paper to polish metal surfaces.

In abrasive wear, a hard particle from one surface indents the other material and forms a groove as material is cut from the other surface. This type of wear is often called gouging wear, as the hard particles 'plough' through the other material. Generally, using harder surfaces improves the resistance to this type of wear. However, there is a limit to how hard the surface can be, since harder materials become more brittle and then the sliding particle can detach large flakes of the other surface. Figure 3.12 shows an example of abrasive wear.

Figure 3.12 Abrasive wear

Abrasive wear is often caused in practice by a contaminant material – perhaps just a hard dust particle – becoming trapped between two surfaces. These particles get embedded in one of the surfaces and are able to abrade the other surface. For example, they can become embedded in the softer of the sliding pair and thus abrade the harder surface.

Abrasive wear can be avoided by selecting materials of similar hardness to work together, and in ensuring 'cleanliness' in the sliding region.

Adhesive wear

Adhesive wear arises when the two surfaces in contact effectively 'weld' together. This occurs by chemical bonding between atoms present in either surface.

How can this happen? There is a very strong driving force – in scientific terms, it is very favourable in terms of the energies involved – for most atoms to form chemical bonds. This is the case in all solid materials under everyday conditions. However, atoms at the surface cannot form their full complement of bonds – they can form bonds with the atoms below them in the solid, but there is nothing outside, if you like. Pressing atoms together in two different surfaces can allow chemical bonds to form.

This is particularly the case for identical materials in contact, and is more likely the cleaner the surfaces. Under a normal load (one acting perpendicularly to the surfaces), the surfaces will deform and, in those regions where the atoms 'bear the load', bonds will be formed. The interface will be no different from a grain boundary: it will have a strength comparable with that of the material itself. In fact, if any plastic deformation occurs in the asperities as they are brought into contact they can work harden and be stronger than the parent material. When these junctions are broken by relative sliding of the surfaces, it will be easier to break the asperity junctions a little way away from the interface itself. As a result pieces of material will be torn out of either of the bodies and left adhering to the other. This wear process is illustrated schematically in Figure 3.13.

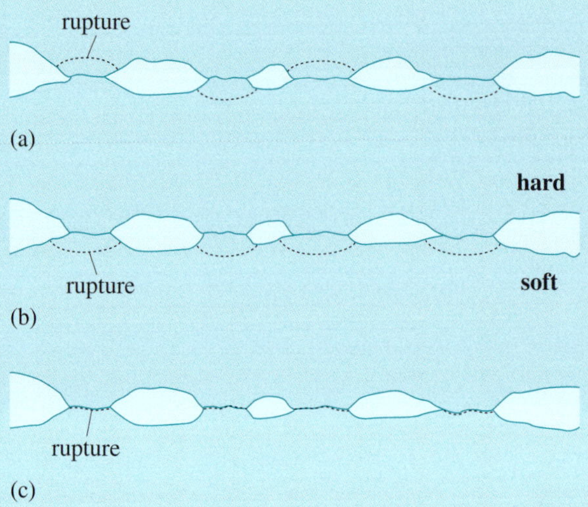

Figure 3.13 A sketch of the contact between clean surfaces and the fracture path taken when the junctions are broken by sliding: (a) similar materials in contact: rupture occurs beneath both surfaces, (b) dissimilar materials: rupture occurs beneath the surface of the softer material and (c) a weak interfacial film: rupture occurs at the interface

At a later stage, these 'picked-up' fragments may be removed by other processes to become wear fragments. The important thing is that when two identical (or similar) materials are rubbed together *both* surfaces become roughened. The asperities on each surface gouge and score the other surface, so that both surfaces deteriorate and rapid wear occurs.

Adhesive wear is more serious between like materials because the tendency to stick to each other is strong. The greatest amount of wear usually occurs in the softer of the two materials in contact. Adhesive wear can be minimized by lubrication which stops the materials bonding to each other.

If adhesion is strong, the combined action of normal and shear forces can cause the surfaces to be drawn into closer contact. This process leads to an increase in the area of contact and, as a result, a point may be reached where the applied force may be unable to shear the enlarged junction. The surfaces are 'locked together' and unless a very much larger tangential force is applied no relative motion occurs between the surfaces. This is the situation referred to as *seizure*.

Fatigue wear

The wear conditions are frequently such that the local stresses around asperities vary with time. This can lead to the slow growth of cracks – fatigue cracks – in the surface material. (Fatigue is considered in more detail in Block 5 Part 4.) These cracks lead to pieces of material becoming detached from the surface. A good example of this situation is in a ball or roller bearing, where fatigue cracks can grow (see Figure 3.14) as the ball or roller cyclically loads and unloads the bearing surface. (Bearings will be described in more detail in Section 3.5.)

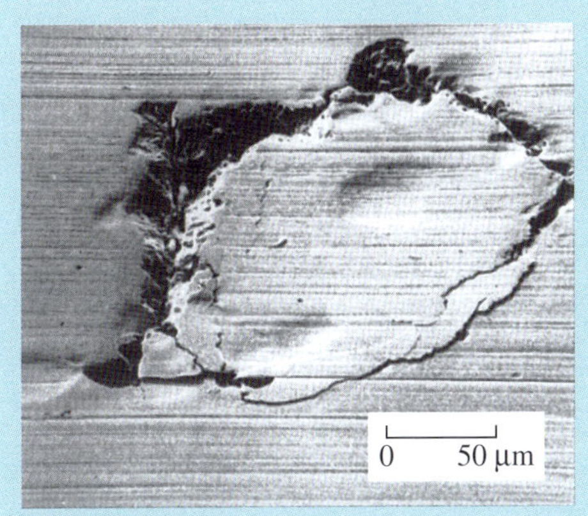

Figure 3.14 Fatigue cracks on the race surface of a roller bearing

Chemical or corrosive wear

Wear processes can also be associated with chemical reactions. The most common example with metals is probably the repeating cycle of formation, removal and reformation of oxides on the surface. The worn fragments here are usually very small, but may appear as compacted flakes comprising a mixture of fine oxide and metal particles. Wear may also occur due to corrosion of a metal surface by, for example, acid species in a lubricant.

Because of its complexity, wear is difficult to express analytically or to quantify. Equations for wear tend thus to be empirical, rather than derived from any underlying theory. One such is:

$$k = \frac{QH}{W} \tag{3.3}$$

This is the Archard wear equation, in which Q is a measured volume of material worn per metre of sliding distance (determined, for example, by weight loss), W is the normal load and H is the hardness of the softer surface. k is known as the wear coefficient and can be used to reflect how much occurs in a particular process.

The measured wear coefficients listed in Table 3.2 give an indication of the severity of wear that is occurring. A value of around 10^{-7} indicates that the wear is very mild, whilst a higher value of around 10^{-3} would indicate that the wear is severe. Generally, the prediction of wear is difficult because the wear rate depends on the way in which material is removed from the surface. Small changes in load or sliding speed, or the environment in which the wear is taking place, can lead to changes in the mechanism (for example, fracture of the surface may occur) and therefore large changes in the wear rate. The wear rate is also dependent on the exact materials that are in sliding contact. For example, in Table 3.2 the wear coefficients are for materials sliding against either tool steel or mild steel. If different materials were chosen then the wear coefficient would change.

Table 3.2 Typical values of the wear coefficient, *k*, measured for dry sliding wear (no lubrication) in tests with a pin of the material sliding against a ring of steel

Material	Type of steel	Wear coefficient, *k*
mild steel	mild steel	7×10^{-3}
ferritic stainless steel	tool steel	1.7×10^{-5}
martensitic stainless steel	mild steel	1.0×10^{-4} *
α–brass	tool steel	1.7×10^{-4}
PTFE, the polymer in Teflon®	tool steel	2.5×10^{-5}
poly(methyl methacrylate), the polymer in Perspex®	tool steel	7×10^{-6}
tungsten carbide	mild steel	1.0×10^{-6}

* reduces to 10^{-9} with lubrication

SAQ 3.2 (Outcome 3.1)

Give one situation where the wear of materials in sliding contact is undesirable and one situation where friction between materials in sliding contact is desirable.

3.3 Lubrication

You've seen that, during sliding contact, it is the deformation and shearing of the surface asperities that give rise to frictional resistance forces and surface wear. If you now introduce a lubricant to the surfaces, you find that the frictional resistance is reduced. With some lubricants, this reduction can be considerable. The lubricant acts by adhering to the asperities and reducing the number of them in potential contact by acting as a sort of cushion. The process of sliding now comprises the deformation and shearing of far fewer enmeshed asperities combined with the shearing of the lubricant substance itself. By definition, the shearing and deformation of the lubricant requires significantly less force (otherwise we wouldn't use it as a lubricant, but as a glue instead!), so overall the frictional resistance of the sliding combination is reduced. This type of lubrication is nowadays called *boundary lubrication*, but used to be called 'greasy lubrication' or even 'dry lubrication'.

Bearings using boundary lubrication go back to the earliest wheels and axles of ancient civilizations such as in China, Sumeria and Egypt. There is also clear evidence that lubricants such as olive oil were applied to aid the movement of large blocks of stone when the Pyramids were being constructed. The lubricants themselves then, and up until the middle of the nineteenth century, were principally of animal, vegetable or marine origin, although Herodotus (484–424 BC) records that oils produced from what we now call crude oil were used for lubrication. Typically, mutton and cattle tallow, pigs' lard oil, neats' foot oil ('neat' is an old word for oxen and cattle), goose grease, fish and whale oils, castor oil, olive oil, cottonseed oil, etc. were used with varying degrees of success. Machines and components such as cart axles, water mills, windmills and potters' wheels were the technology of the times, helping to keep civilization and economies thriving. However, these were all based on relatively imprecisely fitted, slow-moving parts compared to modern-day machines and systems. Classical boundary lubrication was perfectly adequate in all these traditional systems to reduce friction and wear.

It was the Industrial Revolution, and especially the development of the steam engine, the mechanization of mines and the coming of the railways, that spurred on the science of lubrication. The advent of accurately made machine parts and the increase in rotational speeds beyond those of the windmill and cart axle ushered in the concept of *full-fluid-film lubrication*, or *full-film lubrication* for short. In this, the sliding surfaces are completely separated by a lubricant film. There are no asperities in contact at all, and the frictional resistance is solely that of deforming the lubricant itself. That is why it is called full-fluid-film lubrication, and why it dramatically reduces the frictional resistance. Full-film lubrication is a relatively new phenomenon in engineering, although Nature has a head start with fluid-supported hip joints and the like, that you saw earlier. It is the type of lubrication invoked in the Dinorwig plant bearings, and in many industrial applications, car engines, etc.

Fortuitously the discovery of large quantities of mineral oil in Drake's Well at Titusville, Pennsylvania, in the USA in 1859 (see Figure 3.15), ushered in the rapid expansion of the petroleum industry. This included the provision of widely available and highly suitable lubricant oils, just when they were needed.

Figure 3.15 Drake's Well: at the end of August 1859, oil started to flow from this, the first purposely drilled oil well in the world, when the borehole had reached a depth of 21.2 m and just as the financial backers were withdrawing support

---

SAQ 3.3 (Outcome 3.6)

Describe briefly the difference between boundary lubrication and full-film lubrication. Give one everyday example of each.

---

4 WHAT ARE LUBRICANTS?

As you probably know, most lubricants are oils. There are special cases where water or even air may be used in relatively lightly loaded and specialized bearing designs, but the most common engineering applications use oils. Until the last few years, these oils were nearly all extracted from crude oil (i.e. mineral oils), but now we are getting used to more synthetic oils and even vegetable-based oils extracted from non-food agricultural crops. For our purposes, we can group them all together as lubricant oils, eminently suitable for bearings, and we'll have a quick look at why that is.

Exercise 3.3

Think of what a lubricant oil is meant to do, and under what conditions. List as many desirable properties as you can that will help it function as intended.

Mineral oils, synthetic oils (and, up to a point, vegetable-based oils) exhibit all of the characteristics listed in the answer to Exercise 3.3. As you'll see shortly, the low shear resistance is to reduce energy losses and waste heat build-up, and what heat there is is taken up by the oil's own heat capacity. Lubricating oils often have to operate in wide temperature ranges. For example a car engine may sit overnight in a harsh mid-European winter at, say, –20 °C but must be ready for action when it is started the next morning. That same engine and its oil will then be warmed up in a matter of minutes and be running at over 100 °C. It must continue to perform between service intervals of up to about 20 000 kilometres (12 000 miles) which may well include hot summer days and cold winter nights. Oils these days have a lot of additives to satisfy increasing demands on their performance (e.g. higher speeds, longer service intervals), so naturally it helps if they readily take up such additives.

There is little point in a lubricant which attacks the metal bearing surfaces, or softens seal materials in the supply circuits, particularly during longer periods in use, so a good degree of protection or otherwise benevolent behaviour is required. Equally, you wouldn't want to wear special clothing or breathing apparatus every time you opened a can of oil and poured some out. All these attributes are naturally expected at the lowest cost and to be available just about everywhere. The old advertising slogan for a well-known brand of engine oil – 'liquid engineering' – seems truly appropriate.

Lubricants for the Dinorwig plant, although perhaps not requiring quite the same operating characteristics as engine oils, are still subject to arduous conditions.

Before looking at the most important physical property of a fluid lubricant, namely its viscosity, you might want to refresh some basic concepts and properties that relate to fluids in **▼Some things about fluids▲**.

▼Some things about fluids▲

Fluids can be described as substances which yield to any force, however small, tending to alter their shape. More technically, when stationary they cannot sustain a shear force, i.e. a force exerted tangentially to the surface on which it acts. Solids on the other hand can sustain such forces. The concept of fluids includes both liquids and gases (some bearings use air as the lubricant). A liquid has a more or less definite volume of its own, but a gas will normally expand to fill uniformly any container that it is in.

Pressure is defined as the force per unit area exerted in a direction normal to that area. It is usually designated P or p. In SI units, the pressure is expressed in newtons per square metre (N m^{-2}). This unit has been given the special name of the pascal (Pa), so 1 N m^{-2} = 1 Pa. It is equivalent to stress acting on a solid body surface and has the same units. At any given 'point' in a fluid at rest (where the 'point' implies some region that is infinitesimally small but greater than the molecular level), the pressure is the same in all directions.

Density is defined as the mass per unit volume of a substance, and is normally designated ρ (Greek 'rho'). In SI units, density is measured in kg m^{-3}. Density is affected by temperature and pressure, greatly so in the case of gases. For liquids (and solids), the effect is only slight and frequently neglected. In this situation we speak of an *incompressible fluid* (but be warned: this means incompressible compared with a gas – liquids are generally easier to compress than solids). The oil in full-film bearings is usually assumed to be incompressible.

Relative density is the ratio comparison of the densities of two different substances. Quite often the comparative substance is water, in which case the term *specific gravity* is sometimes used.

Specific gravity ironically has nothing to do with gravity! The specific gravity of a substance is the ratio of the density of the substance compared with the density of water. For strict accuracy, the water density is quoted at 4 °C (277 K), when it is a maximum. In bearing design, the relatively small variation of density with temperature is generally ignored. Thus, it is common for the specific gravity of oils to be quoted with both liquids at 15 °C. The specific gravity of most lubricating oils lies between 0.85 and 0.95, so they all float on the surface of water. As it is a ratio, specific gravity has no units.

Specific weight is the weight of a unit volume of a substance, where weight is the force with which a gravitational body (e.g. the Earth itself) attracts the unit volume of substance. As such, therefore, it varies with which gravitational body we are on at the time, and it even varies slightly on Earth. In SI units, specific weight is measured in N m^{-3}.

SAQ 3.4 (Outcome 3.3)

A litre of oil (the same as 0.001 m^3 in basic SI units) has a specific gravity of 0.891 and water has a density of 1000 kg m^{-3}, both at 4 °C. Determine for the oil:

(a) its density;

(b) its mass;

(c) its weight;

(d) its specific weight.

Assume this is on Earth, where $g = 9.81$ m s^{-2}.

SAQ 3.5 (Outcome 3.3)

A simple salad dressing of olive oil and vinegar tends to separate with the oil floating on top. Which has the lower specific gravity?

5 VISCOSITY

The viscosity of a fluid is most simply thought of as its resistance to flow. For a lubricating oil, it is the viscosity that determines the friction loss, heat generation, load-carrying capacity, etc. A freely flowing fluid such as water (or air for that matter) has a low viscosity, whilst a sluggish liquid such as oil is described as having a high viscosity, even though, as you've seen, it is less dense than water.

5.1 Viscosity basics

So what is viscosity exactly?

For a start, we can say what it isn't! It isn't the resistance to static deformation of a fluid, i.e. it is *not* the same as an elastic modulus in a solid.

Let's explore what it is from our base in Wales, at Dinorwig. Figure 3.4 showed you a simplified version of the Dinorwig rotating plant assembly; Figure 3.16 is more of an engineering drawing of the same assembly.

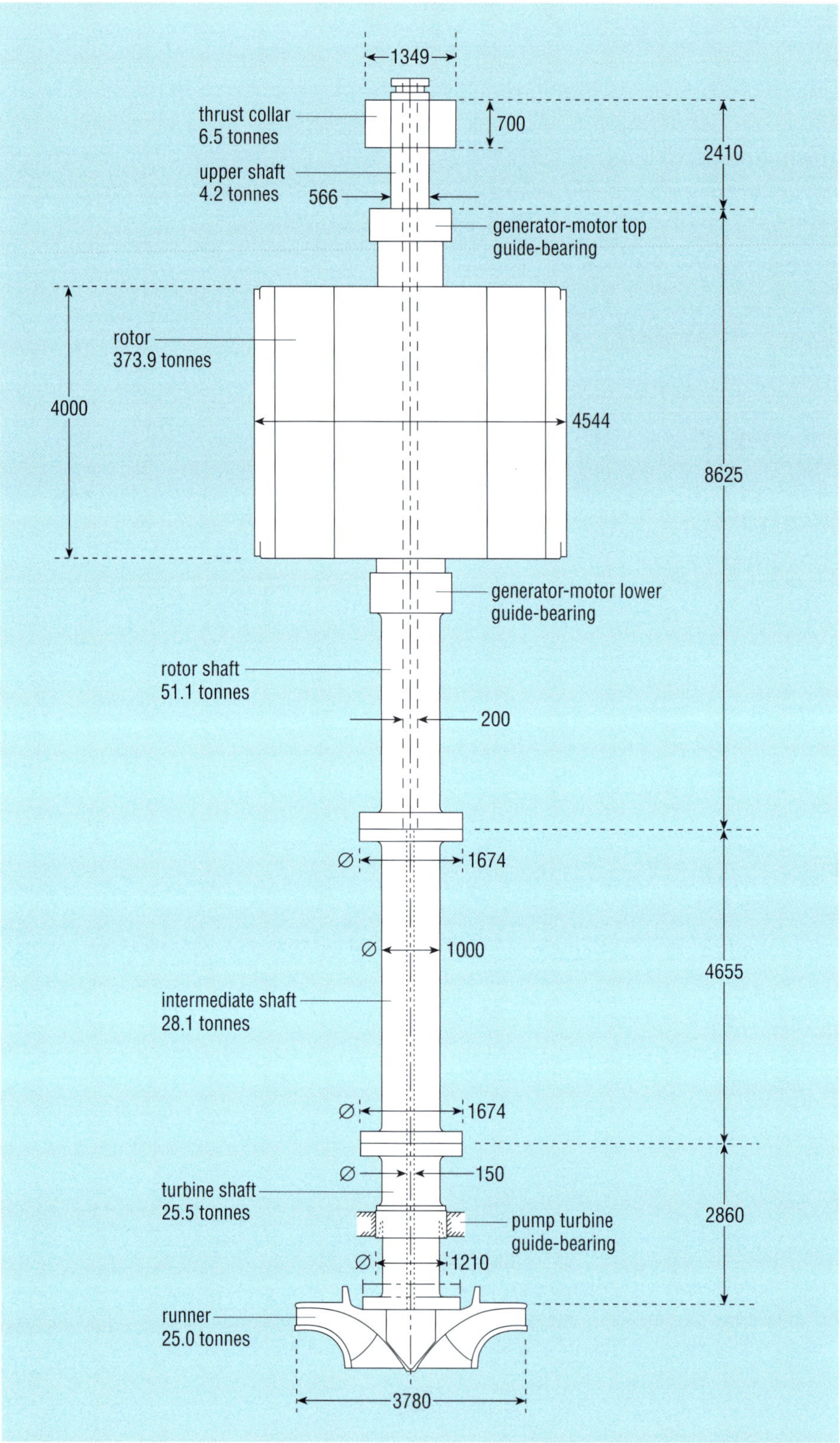

Figure 3.16 Vertical shaft assembly, Dinorwig. All dimensions are in millimetres

Assume that there are films of oil between the rotor and the fixed supports, i.e. within the three journal bearings and the top thrust bearing. Now, if you enlarge small parts of both types of bearing as shown in Figure 3.17, you find you're looking at greatly enlarged portions of the oil film.

In reality, the thickness of the oil film is very small compared to the other dimensions, about 0.1% of the main shaft radius, so you can ignore any effects due to the curvature of the components. A single, enlarged picture of two parallel surfaces is therefore an adequate representation of either bearing,

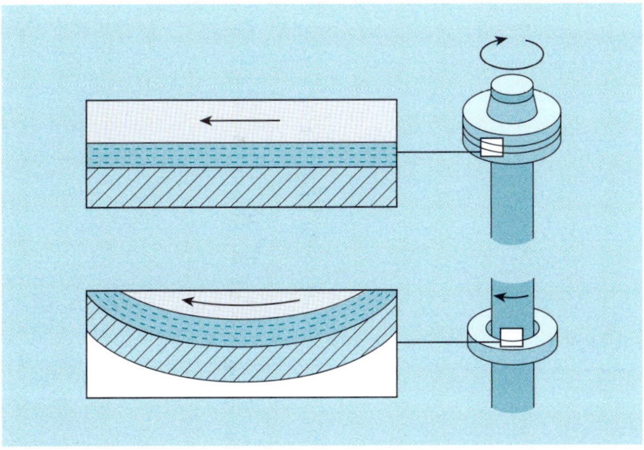

Figure 3.17 Oil films (dotted regions) on thrust and journal bearings

as shown in Figure 3.18. The oil film thickness, *h*, is of the order 0.1 to 0.5 mm for large bearings as in the Dinorwig plant, or about 0.06 mm for the main bearing of the crankshaft in a car engine.

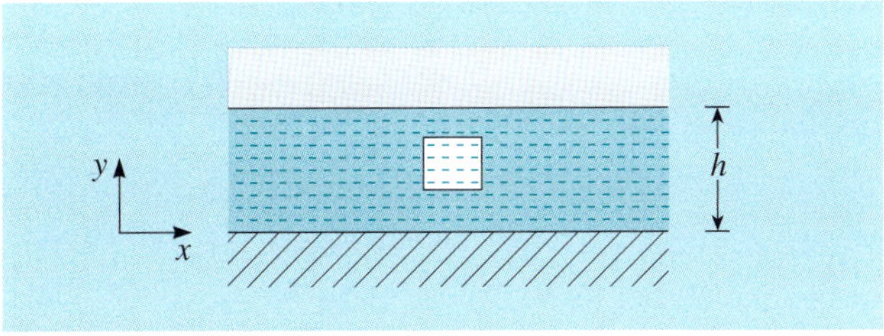

Figure 3.18 Magnified sketch of oil film (dotted region) on bearing, showing element of fluid

Now zoom in closer still and look at a small element of oil somewhere in the middle. Figure 3.19 shows the element enlarged yet again. You have already seen that one characteristic of a fluid is that, *when at rest*, it can't sustain a shear force. Under a load, F, as shown, it would just collapse and spread out.

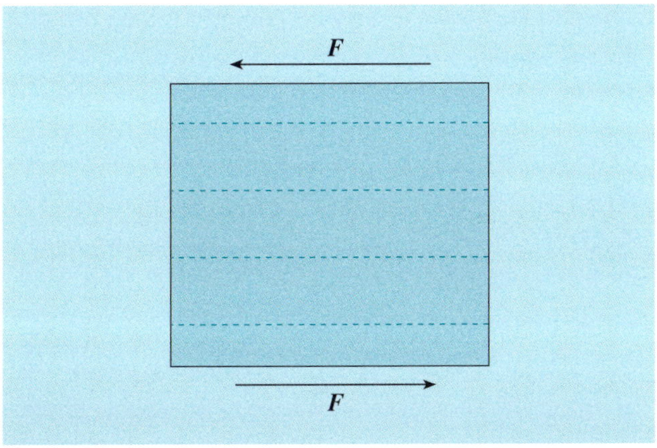

Figure 3.19 The element of fluid from Figure 3.18

Imagine instead a solid, linear elastic material such as metal or ceramic. It *could* sustain such a force. The force F applied to an element of such a solid material would create a shear stress τ applied to the element area, and would distort or strain it by an angle γ, as shown in Figure 3.20. Moreover, when the shear stress was removed the element would return to its original shape, provided that the loading didn't exceed the material's elastic limit.

The element of solid material therefore sustains a shear stress at a corresponding shear strain $\delta x/\delta y$ according to:

$$\begin{aligned}\tau &= G\frac{\delta x}{\delta y} \\ &= G\gamma\end{aligned} \tag{3.4}$$

where τ (Greek 'tau') is the applied shear stress in N m^{-2}, γ (Greek 'gamma') is the shear strain (dimensionless) and G is the shear modulus of the solid material.

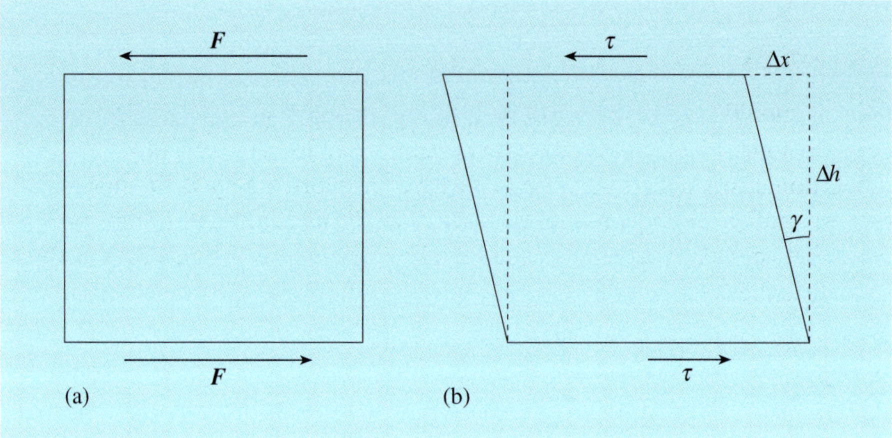

Figure 3.20 An element of solid material subjected to shear

Exercise 3.4

What are the SI units for G?

Now for viscosity. Let's return to the 'parallel-plate model' introduced in Figure 3.18. Imagine the rotating assembly is put into motion so that one plate will have a velocity u and the other will be stationary, as shown in Figure 3.21. In a real fluid, i.e. one that has viscosity, the thin layer of fluid in immediate contact with a solid surface will tend to adhere to that surface. So the oil will adhere to the two parallel plate surfaces: the lowest layer of molecules will remain stationary, the uppermost layer will move with the velocity u of the upper plate. Thus there will be a *velocity gradient* equal to $\delta u/\delta y$ across the oil film thickness. Each intermediate layer of oil will move with a velocity directly proportional to its distance y from the stationary plate. Each layer of oil effectively slides over the adjacent one, an orderly movement known variously as 'streamline', 'laminar' or 'viscous' flow.

For this to happen, there needs to be a force F applied to the moving plate, which generates a shear stress on the element of fluid. In other words, a shear stress is sustained by virtue of the shearing *velocity* – the higher it is, the greater the force. So you can see why it's dangerous to dive into water from too great a height. The faster you enter the water, the greater the viscous force – in the limit with fatal results. Viscosity therefore provides, or sustains, a force not by virtue of a shear strain, but by the *rate* of shear strain. Therein lies the difference between a solid material and a fluid. The resisting force in a solid is *displacement* dependent; in a fluid it is *velocity* dependent.

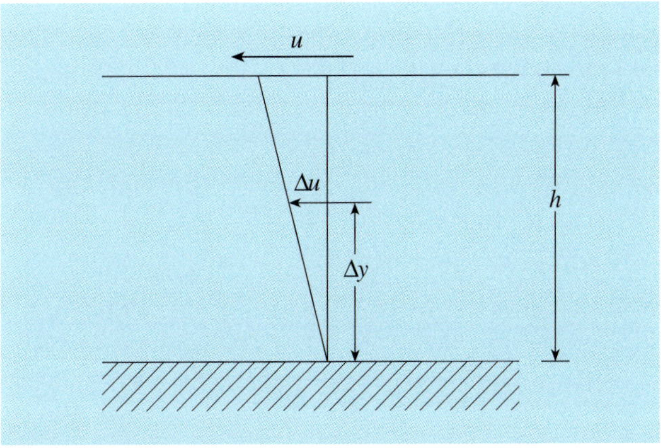

Figure 3.21 Velocity gradient in a fluid film

As you should well know by now, in linear elastic solids the relationship between force and displacement is expressed through Hooke's law: stress is proportional to strain, where the constant of proportionality is the materials property called Young's modulus. Is there an equivalent relationship for fluids between the shear stress and the shear strain rate? The answer is 'Yes', as is described in **▼Newtonian and non-Newtonian fluids▲**.

▼Newtonian and non-Newtonian fluids▲

Newton deduced, after some characteristically thorough experiments and reasoning, that the relationship between the rate of shear strain and the shear stress in fluids was a directly proportional one, i.e.:

shear stress $\propto$ rate of shear strain

If you assign to the constant of proportionality the symbol η (Greek 'eta'), you can write:

shear stress $= \eta \times$ rate of shear strain

$$\tau = \eta \frac{d\gamma}{dt} \qquad (3.5)$$

and η is called the viscosity of the fluid, or more completely the *coefficient of dynamic viscosity*.

The rate of shear strain turns out to be equivalent to the velocity gradient.

Exercise 3.5

If τ is the shear stress, and $d\gamma/dt$ the rate of shear strain, determine the SI units of viscosity.

Note that 'seconds' appear in the units for viscosity. You'd expect this since the shear stress depends on shear *rate*, so time must be involved somewhere.

This model for fluid behaviour, i.e. that the shear stress has a constant relationship to the rate of shear strain, is remarkably accurate for many fluids in real life. Such fluids are called 'Newtonian' fluids and, for our purposes, lubricating oils can be assumed to conform to this behaviour.

It's worth noting, though, that technologically important fluids, especially polymer solutions and melts (i.e. molten polymers), are non-Newtonian. When you shake a bottle of tomato ketchup to make it pour more easily, or you are surprised at how well a thick paint can be brushed onto a surface, you're witnessing what's called 'shear-thinning' – the viscosity of the material falls as a result of it being sheared – not at all what Newton predicted! This phenomenon is very important in the processing of polymers. At the entrances to the mould cavities of injection-moulding tools are narrow apertures called 'gates'. Their purpose is to constrict the flowing polymer melt, which radically increases its shear rate and results in a large drop in the melt viscosity, making it easier to fill the mould cavity.

Thus, in summary, a fluid resists shear deformation by its viscosity, an effect associated with an internal sliding friction between adjacent moving layers of the fluid. The contact layer of a fluid sticks without slip to a stationary surface, or to a moving one, giving rise to velocity gradients within it. A fluid's viscous shear resistance increases with the deformation rate. Making a fluid flow fast requires a greater force than making it flow slowly. Fluid molecules do not go back to their original positions when the applied force is removed. There is no 'strain energy' capability as there is in solid materials. The work done in producing viscous flow is an energy loss transformed on a continuous basis as power absorbed or heat in the fluid, and usually wasted.

5.2 Other viscosities, other units

You've seen that the coefficient of dynamic viscosity, η, is defined by Equation (3.5) and has units of Pa s (pascal seconds). However, there is a range of other units and ways in which viscosity is described. One commonly used one, that you may have come across, is in oil specifications that quote viscosity in centistokes (cSt).

So what's that all about?

This is a measure of *kinematic viscosity*, which is simply the dynamic viscosity divided by the fluid density, ρ. It's given the symbol ν (Greek 'nu') so:

$$\nu = \frac{\eta}{\rho} \tag{3.6}$$

It is used because the term η/ρ frequently crops up in fluid mechanics, and is a quantity that can be determined directly from simple measurements.

Exercise 3.6

By inspection of Equation (3.6), determine the SI units for kinematic viscosity. (Hint: from Newton's Second Law of Motion, $1\ \text{N} = 1\ \text{kg} \times 1\ \text{m s}^{-2}$.)

You can see from the answer to Exercise 3.6 that the units of ν contain no forces, or masses, as these have cancelled out. What remain are units of position and time, which is the field of kinematics; hence the name 'kinematic viscosity'. Kinematic viscosity effectively relates shear characteristics (dynamic viscosity) and fluid acceleration (mass density) so perhaps can be thought of as a more complete measure of the 'flow' property of a fluid.

One typical problem that engineers have to deal with is that, for historical reasons, kinematic viscosity is still widely quoted in cgs units, i.e. ν is in $cm^2\ s^{-1}$, which is given the name 'stoke' (symbol St). Worse still, to make the numbers easier to handle the stoke is subdivided into centistokes, where 1 cSt = 0.01 stoke. So 1 stoke is 1 centimetre squared per second, and 1 centistoke is one-hundredth of this. Thus 1 cSt = $1.0 \times 10^{-6}\ m^2\ s^{-1}$. Confusing isn't it? (And I haven't even mentioned the imperial units still widely used in the USA!)

If you ever have anything at all to do with lubrication and viscosity, this issue of units is one that you will have to be familiar with. It is worth a bit of effort to wrestle with, however, because it classically illustrates what engineers often have to deal with in real life – coping with obsolete measurements, despite the widespread adoption of SI units. It will also help a bit later in the text.

SAQ 3.6 (Outcome 3.4)

A litre of oil has a quoted kinematic viscosity of $\nu = 32$ cSt and a density of 891 kg m^{-3}. Using the data given in SAQ 3.4 (page 158), determine its coefficient of dynamic viscosity.

(The conversions in SAQ 3.6 might seem a bit confusing. It's an unfortunate fact that the use of SI units often results in awkward numerical values – that's why many people cling to the older systems. Despite this, you're still strongly recommended to stick with the SI system. In particular, this avoids the complications of irregular conversion factors, which can be a potent source of error in other systems of units.)

5.3 Viscosity grade system for oils

The International Organization for Standardization (ISO) system specification for viscosity is based on grade ratings for the kinematic viscosity of the oil, in centistokes, measured at 40 °C. Each viscosity grade is numerically equal to the kinematic viscosity with an allowed variation of ±10%. For example, the oil in SAQ 3.6 would be called ISO VG 32 (ISO viscosity grade 32) which would have a kinematic viscosity of 32 cSt at 40 °C, with a tolerance of ±3.2 cSt. The oil could thus have a kinematic viscosity of anywhere between 28.8 and 35.2 cSt at 40 °C. As it happens VG 32 oil is quite common, and would be equivalent to an engine oil, whereas VG 150 oil is a thicker (more viscous) gear oil. The larger the grade number, the more viscous the fluid.

SAQ 3.7 (Outcome 3.5)

How would you grade a litre of oil whose coefficient of dynamic viscosity at 40 °C was 0.0891 Pa s and whose density is 891 kg m^{-3}?

5.4 Temperature and pressure effects on viscosity

An important issue with both mineral and synthetic oils is the effect of temperature and pressure. Having seen that viscosity is a constant (of proportionality) in Equation (3.5), you have to realize that it is nonetheless highly influenced by temperature and, to a lesser extent, by pressure. The viscosity of mineral and synthetic oils decreases markedly with increasing temperature, as the increased thermal energy enables their molecules to escape from their neighbours' influence more easily. Figure 3.22 shows the variation for a typical oil. In the working range of temperatures we can normally expect in most bearings (i.e. say 0 to 110 °C), the viscosity can fall with rising temperature by a factor of a hundred or more.

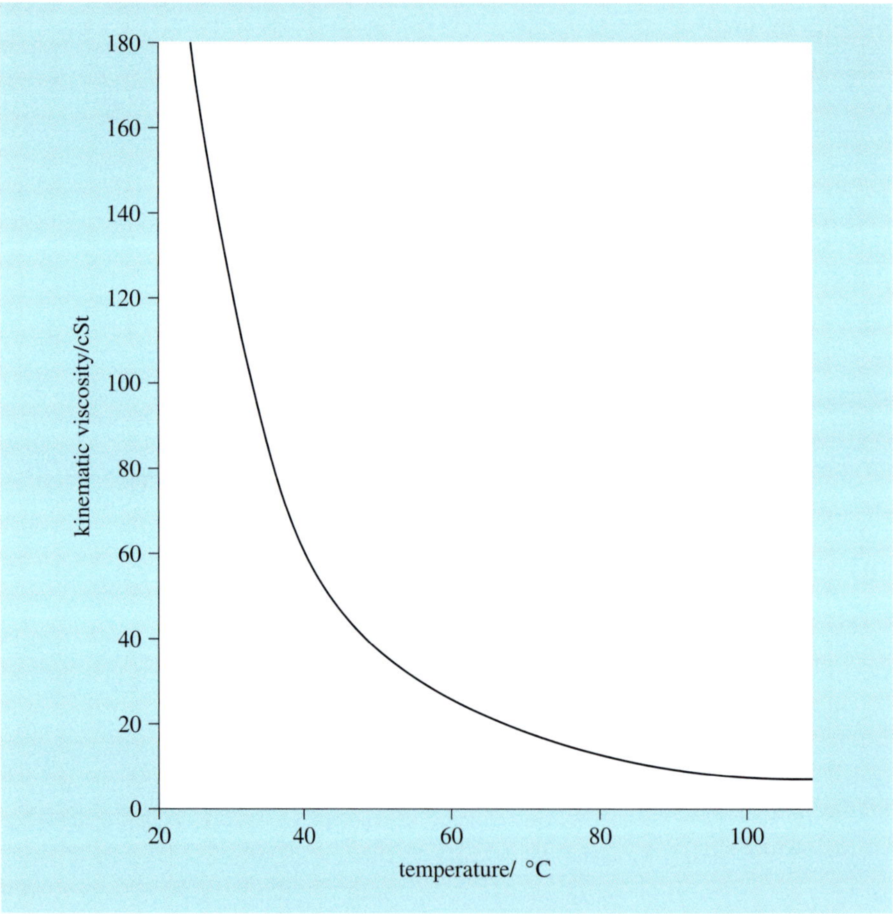

Figure 3.22 The variation of viscosity with temperature for a typical lubricating oil

This can be a disadvantage or an advantage, depending on the circumstances. For example, in the case of a car engine oil, too high a viscosity when the engine is started cold would mean excessive drag causing a heavy drain on the battery. More importantly, a few seconds' delay before full-film lubrication is established in critical areas each time the engine first fires could cause excessive wear, shortening the engine life. Then when the engine is hot, the oil film may be too thin to develop full-film lubrication in the same critical areas. Both of these are obvious disadvantages. On the other hand, with the engine warmed up, the reduced viscosity reduces energy losses, as you'll see in the next section. This is of course an advantage.

The effect of pressure is the opposite to that of temperature. As pressure increases the viscosity also increases. In other words, the greater the 'squeeze' force, the higher will be the viscosity and the greater the resistance to the fluid being squeezed out. This can be a fortunate state of affairs in extreme conditions, for example, between gear teeth – the harder the lubricant is squeezed, the more viscous it becomes, thereby delaying the chance of surface-to-surface contact. Also, it often happens that the extreme pressure or load region occurs where the temperature is highest, so a reduction in viscosity due to the temperature increase is (partially) offset by an increase in viscosity due to the pressure increase. For most situations in full-film lubrication, however, in particular for the bearings and conditions we'll be looking at, the effect of pressure on viscosity is small enough to be neglected. We are trying to avoid excessively high pressures and it becomes an issue only at very high pressures.

Perhaps an obvious question to ask here is, 'How does all this affect the oil that is put into my car engine?' See ▼**So what does 10W40 mean?**▲.

▼So what does 10W40 mean?▲

When you buy oil for you car's engine, you'll find it labelled something like 'Multigrade, 10W40'. This is a designation in a viscosity grading system known as SAE after the Society of Automotive Engineers in the USA. Unlike the ISO system, numbers in the SAE system are not the viscosities in centistokes at 40 °C, although the higher the number, the higher the viscosity. In the SAE system, both the dynamic and kinematic viscosities are used, and two reference temperatures 100 °C and 40 °C are used in the specification. The grades followed by the letter 'W' are winter grades, designed for use at lower temperatures. The W-classification is based on a maximum low-temperature dynamic viscosity, a minimum kinematic viscosity at 100 °C, and a maximum borderline pumping temperature. Thus a 10W grade will have a dynamic viscosity that does not exceed 3500 centipoises (= 3.5 Pa s) at –20 °C, a maximum borderline pumping temperature of –25 °C and a minimum kinematic viscosity at 100 °C of 4.1 cSt. The non-W-classification is a bit simpler. A 40-grade has a minimum kinematic viscosity of 12.5 cSt and a maximum of less than 16.3 cSt, both at 100 °C.

6 TYPES OF BEARING

The Dinorwig plant represents a demanding and severe set of conditions for any bearing – high speeds, large loads, changes in loads, minimum energy loss and, perhaps most important, utter reliability day in, day out. So what choices of bearing type are there?

There are many engineering components which involve the rubbing – or potential rubbing – of two surfaces. Industrially, the most important is the enormous field of bearings. A 'bearing' is a generic term used to describe any support device for keeping a moving (usually rotating) part in position. Three types of bearing are shown schematically in Figure 3.23, two of them for rotation and the third for the sliding of one surface against another.

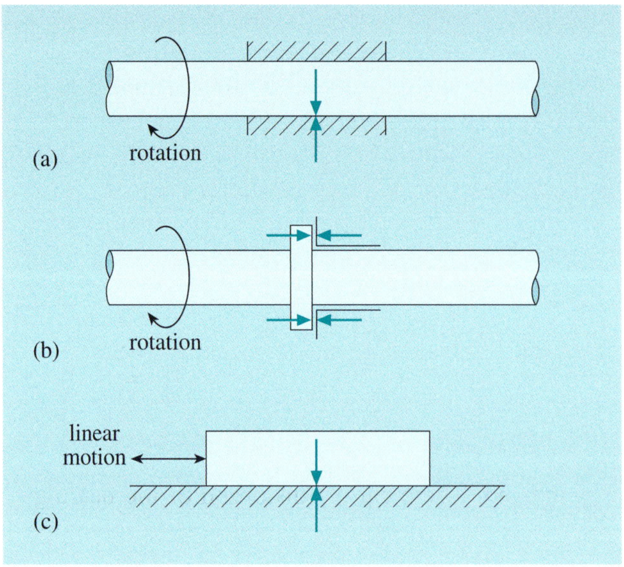

Figure 3.23 (a) A journal bearing, where load can be transmitted radially from the shaft as shown. (b) A journal bearing that has axial loads. (c) A linear sliding bearing

In most such applications, lubricants are used in order to maintain separation of the surfaces and prevent contact. Apart from simply putting a lubricant between two sliding surfaces, to form either boundary or full-film lubrication, there are other forms of bearing that you may be familiar with.

Probably the most common are rolling element bearings, or as often abbreviated, rolling bearings. In these, a number of elements (e.g. balls or cylindrical pins), literally roll between the two surfaces carrying the load. The balls or pins are usually held in position by a so-called 'cage' which keeps them apart and ensures that they are held correctly between the support surfaces. The rolling elements and their support surfaces ('races') are usually made of a high-grade steel and are finish-machined to high precision. Familiar examples of rolling element bearings are the ball bearings in the fork tubes and the bottom bracket spindles of bicycles. They are also widely used in industry and in cars, especially in gearboxes, and can use either oil or grease as a lubricant.

Rolling bearings offer high accuracy and low frictional resistance but can be adversely affected by any debris particles or contaminants in the lubricant. On the scale needed for the Dinorwig plant, they would be large, heavy and hideously expensive. More importantly, their lives are ultimately limited, even when correctly installed and used, by fatigue of the load-bearing balls or pins (remember Figure 3.14). As each ball or pin rolls between the races, the material is loaded and unloaded as it passes the contact point. Because the materials at the contact points deform very slightly under load they are subject to cyclic strains and stresses as the bearing operates, and the stress pattern will include tensile ones, a classic fatigue scenario. Ultimately then, the bearing has a design life in terms of revolutions in use, however many millions that may be.

Broadly speaking, work on bearings falls into two camps. One is predominantly interested in the science of materials rubbing together, such as in boundary lubrication, or perhaps investigating new combinations of materials that can offer low frictional resistance. This may be important in areas where the presence of lubricants such as oil or grease is not allowed. This camp tends to be in the domain of materials scientists and engineers. The other camp is more interested in keeping the surfaces apart at all times – full-film lubrication – and is much more in the domain of fluid mechanics. Here, surface-to-surface contact is viewed as a mode of failure, so perhaps it is natural for there to be some sort of divide between the two camps. In reality, of course, there's an overlap in knowledge and skills, if only because the surfaces in a full-film lubrication bearing will inevitably touch at times.

If operating conditions in a bearing were ideal, there would be no problem in the selection of a bearing material. For example, if the surfaces in a bearing were always separated by a film of lubricant, if there were no elastic or thermal distortions, no roughness of the surfaces, and no grit or other abrasive particles in the lubricant, almost anything would serve as a bearing material. The only consideration would be to select a material of adequate strength. Steel bearings would probably be standard. However, ideal conditions are almost never attained. There can be distortion of the bearing shell and its housing. The shaft might deflect slightly and tend to rub at the ends of the bearing. Bearing and shaft surfaces might not be as smooth as they should be, allowing some peaks of surface roughness to penetrate the lubricant film. Conditions of operation, such as starting or stopping or when the shaft is rotating at very low speeds, imply that fluid films are not developed enough to separate the rubbing surfaces. Consequently various mechanical, metallurgical and chemical characteristics of bearing materials are necessary to augment performance for the times when lubrication is not fully developed. Some of these requirements are explored in **▼Bearing materials▲**.

▼Bearing materials▲

In general, the standard requirements for bearing materials revolve around the following properties:

score resistance;

compressive strength;

deformability;

fatigue strength;

embeddability;

corrosion resistance;

structure; and last, but by no means least,

cost.

No single composition has been found which will have all the above characteristics equally. For example, high fatigue strength generally means poor deformability. No order of importance is implied in the above listing. For example, in damp environments, corrosion resistance may be of the utmost importance, whereas in industrial environments embeddability may be the most significant. Let's look at the requirements in a little more detail.

A satisfactory bearing material should not damage the surface of the journal with which it is used even when operating under conditions of boundary lubrication or, in extreme cases, when dry. Some materials exhibit a tendency to adhere or weld to the shaft if fluid film conditions break down. The **score resistance** of a bearing material is a measure of this antiweld or antiseizure characteristic. If the score resistance is not sufficiently high, the bearing may score, or 'wipe', on the shaft. If operating conditions are sufficiently severe, the shaft may even seize within the bearing causing severe damage.

Compressive strength is the ability of the bearing material to carry the imposed load without distorting, for example by extruding out of shape. However, some deformability might be required (see below).

Deformability is the ability of a bearing material to yield to deformation while operating, without causing failure, and could be highly important. For example, upon first use (because of the limits of machining accuracies), or with a new loading condition, if the bearing material is deformable to a degree it will adjust itself to suit by wearing or wiping without causing a serious high-temperature condition to develop.

If the load is variable in magnitude or direction or both, the corresponding variable stresses imposed upon the bearing material introduce conditions that may invite fatigue. For example, engine bearings such as main and connecting-rod bearings of crankshafts carry high alternating loads, and so their **fatigue strength** needs to be high. As in other structures, the fatigue life of the material depends upon the range of stress to which it is subjected and the number of stress cycles.

The **embeddability** characteristic is important when dirt enters the clearance space. The dirt could either embed itself in a relatively soft bearing material or jam itself into the clearance space between a harder bearing material and the shaft. Such dirt can cut or gouge a groove in the bearing and shaft. Hard metal particles and other foreign material can introduce this difficulty, and will almost certainly occur at some time or other, particularly after repairs, maintenance and even on first assembly.

The requirement for engine bearing materials to be **corrosion resistant** came sharply into focus in the 1930s when high automobile sump temperatures caused an epidemic of oil oxidation and deterioration and the formation of acid. This acid proceeded to attack vigorously and effectively some of the then-newer bearing materials, with disastrous results.

For each type of material there is probably a particular metallurgical **structure** which gives the best performance. In metals, it was once thought that a good bearing metal had to consist of a duplex structure with hard particles supported in a soft matrix. Another wide class of bearing alloys consists of a continuous matrix of a harder material with a small amount of a softer material finely dispersed through it. The supporting structure must of course be sufficiently strong and stiff to carry loads further into the component.

The selection of a proper bearing material may be influenced by **cost** and by the availability of the material itself. The cost of a single bearing may be small as compared to the total cost of the machine. For example, the Dinorwig plant is a huge capital investment and the bearings are a relatively low-cost item, but in real terms they would still cost well into five-figure sums each to replace.

So what materials make good bearings in the real situation? As you might expect, this problem was of prime importance in the developing steam age when full-film lubrication came to the fore, and is just as relevant today. In 1839, Isaac Babbitt obtained a US patent for a special bearing alloy. This was largely tin, but contained small amounts of antimony, copper and lead (Leonardo da Vinci, in the 1490s, wrote about a low-friction alloy for bearings made from 'three parts of copper and seven of tin melted together', but knowledge of this was lost until his Codex Madrid was found in 1967.) Babbitt's alloy was used as a liner in a journal bearing and proved to be an excellent bearing material. It had a silvery appearance and would now generally be described as 'white metal' or as 'Babbitt metal'. For many decades since, white metals have been the mainstay of bearing design and have been refined and developed to be the subject of specification standards. Broadly speaking, there are two types: tin-rich and lead-rich. It is usual for the white metal to be 1 or 2 mm thick, cast *in situ* within a steel bearing housing and finish-machined (for size and surface finish). A typical composition of a tin-rich white metal might be:

antimony 7.0–8.0%

copper 3.0–4.0%

lead 0.35%

cadmium 0.05%

arsenic 0.1%

zinc 0.005%

bismuth 0.08%

iron 0.1%

aluminium 0.005%

others 0.2%

remainder (approx. 88%) tin.

SAQ 3.8 (Outcome 3.9)

Briefly explain the main components of white metal alloys and why they are necessary in bearings.

7 KINETIC ENERGY AT DINORWIG

At Dinorwig, the 510-tonne shaft assembly rotates at 500 rev min^{-1}, supported by the film of oil in the thrust bearing and located laterally by the films of oil in the journal bearings. The oil films in the clearance spaces are only fractions of a millimetre thick. These thin films of oil in the bearings are all that allow the massive rotating assembly to function for hours on end. They are absolutely vital. If for some reason they break down or the oil supply gets cut off, there will be surface-to-surface contact. In the worst-case scenario, this will lead to a rapid build-up of heat, seizure, sudden release of energy as explosive failure, and disintegration with massive heavy parts flying about the place. This could all happen in a few seconds, too quickly for personnel to run away from the scene – clearly a completely undesirable situation, whether at Dinorwig or any other industrial plant.

It is worth a bit more thought about the scale of the problem and the energies involved. Common practice in engineering is to carry out approximate calculations (often called 'ballpark' estimates, or 'back-of-the-envelope' sums) to get a feel for the numerical values of various parameters and their influence on performance or other predictions. These are useful, for example, in verifying predictions of computer models further down the line. So what you have is a complex rotor and shaft assembly rotating about the longitudinal vertical axis (see Figure 3.16). How can you model this simply?

The first thing you need to know is that for a rotating shaft the kinetic energy can be expressed as:

$$\text{kinetic energy} = \tfrac{1}{2} I \omega^2 \tag{3.7}$$

Perhaps confusingly, *I* is also used as the symbol for second moment of *area*, measured in m^4.

where I is the second moment of mass (sometimes 'moment of inertia') in kg m^2, and ω (Greek 'omega') is the angular speed in rad s^{-1} (radians per second), both about the axis of rotation. The angular speed is easy enough – we just convert the quoted speed in revolutions per minute to radians per second, in fact you can do this in a short while.

The second moment of mass of a solid disc or shaft is given by:

$$I = \tfrac{1}{2} m R^2 \tag{3.8}$$

where m is the mass in kg and R is the outer radius in metres. I therefore depends on the mass and, more importantly, on how that mass is distributed. A thin, wide disc has a higher second moment of mass than a thicker, less wide one of the same mass. The effect of the distribution, namely the radius, is squared, so if you want a high value for I, it is best to arrange for the mass to be as far away from the axis of rotation as possible. Conversely, for a low value of I, you want the mass to be as close to the axis as possible. At Dinorwig, you are stuck with the dimensions as they are, because they already exist and were determined by a range of factors including performance, strength and stiffness.

For a hollow shaft or disc, the second moment of mass I is:

$$I = \tfrac{1}{2} m \left(R^2 + r^2 \right) \tag{3.9}$$

where r is the inner radius in metres, m is the mass of the hollow shaft and R the outer radius as before. You might think that the '+' should be '–' as you are effectively extracting the effect of the missing inner hollow portion. In fact, it works out this way because the mass m is now the mass of the remaining solid bit of material. If you first work out I for the solid shaft using its mass, and then I for the extracted shaft forming the hollow with its mass, you end up with a net I for the remaining hollow shaft as in Equation (3.9). So now you can calculate the second moment of mass of the rotating assembly. Simply consider the second moments of mass of all the different disc and shaft portions and sum them up.

You need to introduce a few approximations at this stage. This is a perfectly acceptable procedure, but you must remember what you've done and the assumptions made in subsequent discussions or analysis. For example, the turbine runner is a complicated shaped piece. As you know its mass and outer radius, it's reasonable for the purposes of this exercise to consider it as a solid disc of the same mass and radius. The result won't be a million miles out from the true situation.

SAQ 3.9 asks you to have a go at determining the second moment of mass of the rotor on its own.

SAQ 3.9 (Outcome 3.2)

The mass of the rotor shaft including the bits sticking out for the bearings and couplings is 51 tonnes. The mass of the rotor is 374 tonnes. The outer diameter of the rotor shaft is 1.000 m passing through the rotor; the shaft is hollow with an inner diameter of 0.200 m. The outer diameter of the rotor is 4.554 m.

First decide which would be expected to have the higher second moment of mass – the rotor or the shaft – and then use the appropriate formulae to determine the second moments of mass of the rotor and the rotor shaft.

Table 3.3 lists my estimates for m and I based on the dimensions and data on Figure 3.16 for all the component parts of the assembly. The totals for the whole assembly are then given by simple addition. Also shown are the relative contributions of each piece to the total set-up, which makes interesting reading.

Table 3.3 Dynamic data for the Dinorwig shaft assembly

Component	Mass / 10^3 kg	% of total mass	Second moment of mass, I / 10^3 kg m^2	% of total I
Thrust collar	6.5	1.4	1.7	0.2
Upper shaft	4.2	0.8	0.2	< 0.05
Rotor shaft	51.1	9.9	6.6	0.6
Rotor unit	373.9	72.7	1016.0	94.3
Intermediate shaft	28.1	5.5	3.6	0.3
Turbine shaft	25.5	4.9	4.7	0.4
Runner	25.0	4.9	44.5	4.1
Totals	514.3		1077.3	

The specifications for Dinorwig quote an actual mass of just under 510 tonnes for a complete rotating assembly, and the quoted value of I for the entire generator rotor (which includes the shaft and the more complex shape of the rotor assembly) is 1030.5×10^3 kg m^2. The estimates in Table 3.3 are therefore within a per cent or so – not at all bad for 'ballpark' calculations. Remember that the whole point of doing this exercise was to estimate the kinetic energy of the rotating parts in normal operation. Now you can have a go at this.

SAQ 3.10 (Outcome 3.2)

From the data now evaluated determine the kinetic energy of the rotating parts of one of the Dinorwig generating sets, running at 500 revolutions per minute.

SAQ 3.11 (Outcome 3.2)

How high could an average car (say 1.5 tonnes) be raised from the ground using the Dinorwig rotor assembly kinetic energy calculated in SAQ 3.10? Have a guess at the answer first.

My guess beforehand was that it would be something like a few hundred metres, 'blowing' the car to the top of the Eiffel Tower for example, but 70 miles! That's a low Earth orbit. I found this figure so large that I first thought I must have made a mistake somewhere, like introducing an extra 10 or even 1000 in my sums. But no, this much energy is indeed contained in one rotor assembly when operating. And there are six of them – thank goodness it's all inside a mountain!

That is a lot of energy to be safely contained by the support bearings, even when the generator set is not producing any power or pumping back any water. Remember that it is the kinetic energy contained by the rotating parts themselves which would need to be dissipated if, for example, anything went wrong and a speedy shutdown was needed. Over and above this rotating energy, of course, is the work done by the system in generating power or pumping the water back.

The bearings supporting this rotating mass have, therefore, above all else to be utterly reliable under seriously demanding loading conditions. They should also be efficient in terms of reducing any power absorbed and consequent energy loss. Fluid-film bearings are both reliable and efficient, and, when working correctly, ensure that the moving and stationary support surfaces are free of all contact. The surfaces are thus separated by a thin film of fluid – the lubricant itself. But how efficient? Can the losses be quantified?

8 ENERGY LOSSES IN FULL-FILM LUBRICATION

Even with running surfaces completely separated by an oil film, there will be an energy loss in bearings using full-film lubrication because of the oil's viscosity. Figure 3.24 is a simplified representation of one of the Dinorwig vertical shaft journal bearings. To start with, let's look at a lightly loaded scenario with the shaft (journal) running concentrically in the stationary bearing housing. Remember that in reality the oil film thickness is very small, of the order 0.5 mm, so Figure 3.24 greatly exaggerates the scale of the oil film. This gap for the concentric arrangement is called the *radial clearance*.

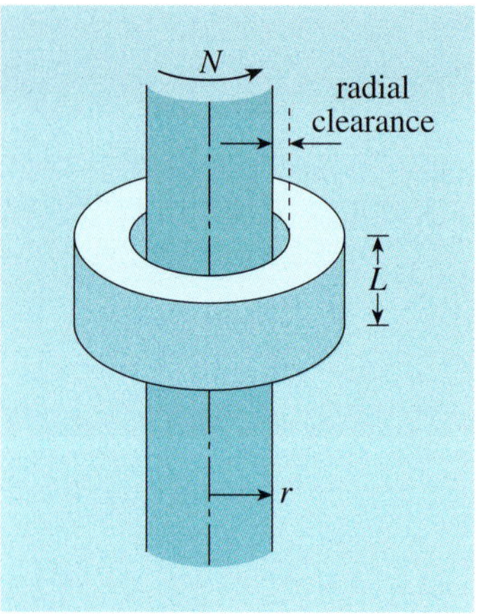

Figure 3.24 A simplified sketch of a journal bearing

What is required is an estimate of the *rate* of energy loss, i.e. the power absorbed, while the bearing is operating. To do this, you first need to estimate the shear stresses acting in the oil film, then the shear force, then the shear drag torque due to this shear force, then the power absorbed due to this torque. It's useful, as before, to start with some simplifying assumptions. Because the oil film is very thin relative to the radius of the bearing, you can neglect the effect of the curvature of the bearing surface. Thus the oil film may be treated as an unwrapped body having a thickness equal to the radial clearance and with the moving layer of oil 'stuck' to the surface of the shaft, in comparison with the stationary layer 'stuck' to the surface of the bearing. You can also assume that there's no flow leakage of oil axially along the shaft out of the bearing (in reality there is, of course – there's nothing to stop it coming out).

The main data needed are:

shaft speed $= N$ rev sec^{-1} or $\omega = 2\pi N$ radians s^{-1}

shaft radius $= r$ m

bearing axial length $= L$ m

radial clearance $= C_r$ m

dynamic viscosity $= \eta$ Pa s

The power absorbed can be estimated in four steps.

Step 1: the shear stress in the oil film

Using the Newtonian model:

shear stress acting in the oil film = rate of shear × viscosity

The maximum rate of shear is described by the velocity gradient:

$$\frac{du}{dy}$$

where

$$\frac{du}{dy} = \frac{\omega r}{C_r} \quad (3.10)$$

(neglecting the small changes in r due to the film thickness).

So the shear stress acting is:

$$\tau = \frac{\omega r}{C_r}\eta \quad (3.11)$$

Step 2: the shear force acting in the oil film

The shear force is simply the shear stress times the area on which it acts:

$$F = \frac{\omega r \eta}{C_r} \times 2\pi r L \qquad (3.12)$$

(This assumes that the fluid film extends right round the bearing.)

Step 3: the shear drag torque due to the shear force

The shear drag torque is the shear force times the shaft radius (again neglecting small changes in r):

$$\begin{aligned} \Gamma &= \frac{\omega r \eta}{C_r} \times 2\pi r L \times r \\ &= \frac{2\pi \eta r^3 L \omega}{C_r} \end{aligned} \qquad (3.13)$$

So what this equation gives you is the amount of torque needed to keep a bearing turning at a given speed against the resistance due to the viscosity of the lubricant. The other quantities involved are simply the dimensions of the bearing and the size of the gap between the rotating and stationary components. This important tool for bearing engineers is called *Petrov's equation* after Nikola Petrov, the Russian who published it in 1883. Petrov was a railway engineer, who was also an outstanding applied scientist.

Step 4: the power absorbed due to the torque

It follows therefore that the power loss P associated with this drag torque when the shaft speed is ω rad s^{-1} is given by:

$$\begin{aligned} P &= \Gamma \omega \\ &= \frac{2\pi \eta r^3 L \omega^2}{C_r} \end{aligned} \qquad (3.14)$$

This can be checked by considering the units:

$$\frac{\text{N s}}{\text{m}^2} \times \frac{\text{m}^3 \times \text{m}}{\text{m} \times \text{s}^2} = \text{N m s}^{-1} = \text{J s}^{-1} = \text{W} \qquad (3.15)$$

The total energy lost for a period of running will be simply the power absorbed times the time of operation:

$$\text{J s}^{-1} \times \text{s} = \text{joules} \qquad (3.16)$$

By studying Petrov's equation, or the power-loss expression, you can see the effect of the various parameters on energy losses.

SAQ 3.12 (Outcome 3.7)

Make a list of the parameters and their effects on the power absorbed in running the bearing. Given a free hand, what could you suggest changing to reduce the power absorbed, i.e. to reduce the total energy lost?

You can see from the answer to SAQ 3.12 that reducing both the shaft size and its speed would be highly advantageous – their effects are greater than linear. Increasing the film thickness and decreasing the bearing length have linear effects, as does decreasing the viscosity. There is not always the freedom to change things as one would wish, however, as there are usually other constraints to consider. For example, it may not be safe to decrease the viscosity too much as there may not be enough support if the loads increase. Equally, the bearing clearance and length are set by other factors not yet considered.

It is now your turn to have a go at applying Petrov's equation to one of the Dinorwig plant bearings. Take the pump/turbine guide bearing, journal bearing 3 in Figure 3.4, the lowest one in Figure 3.16. The parameters of the bearing are as follows:

journal diameter = 1.210 m

journal speed = 500 rev min^{-1}

bearing axial length = 0.350 m

radial clearance = between 0.3 and 0.4 mm

oil viscosity = VG68

oil density = 900 kg m^{-3}

SAQ 3.13 (Outcome 3.7)

Using Petrov's equation, determine the maximum torque required to operate the bearing at 40 °C. (*Caution:* Take care with the units!) Determine also the power absorbed, and the total energy lost after one hour's running.

It is interesting how much power is absorbed by the bearing and the consequent energy loss. Going back to our 1.5-tonne car and its confused occupants, one hour's running is equivalent to lifting the car 100 km! This is just for one bearing – there are three journal bearings and one large thrust bearing for each plant set, and there are six of these plant sets. It would seem therefore that there is ample scope for reducing such energy losses.

But how?

As you will know by now, engineers don't always have the luxury of choosing what problems to solve, or a 'clean sheet of paper' to start from. Improvements often have to start from where we're at. For example, let's assume that our bearing's physical dimensions, r, L, C_r, can't be altered, neither can the speed, N. That seems to leave only viscosity, η, with which to play if we want to keep the power losses down; or does it?

It is good discipline in engineering when confronted with situations like this to go back to first principles and double-check all the assumptions that are made in developing our models of understanding. Let's do this now.

SAQ 3.14 (Outcome 3.6)

Go back to the derivation of Petrov's equation and study the steps involved and any assumptions at each stage. Suggest how it might be possible to reduce power losses in a full-film journal bearing by, say 20%, if you can't alter r, L, C_r, N or η.

This is precisely what can be done – put gaps in so that the wetted bearing surface does not cover the full circumference. Naturally there is a bit more to it than this, but congratulations if you spotted the option.

Now let's come to the crucial issue of how the bearings actually work in carrying the loads.

9 HYDRODYNAMIC LUBRICATION

9.1 What *is* hydrodynamic lubrication?

So far it's been assumed that the shaft (also known as the journal), whether or not it is loaded, is separated completely from the bearing surface by a film of oil. The power-loss calculations of Petrov's equation have been based on the fact that the shaft is perfectly concentric within the bearing. In other words, the fluid film thickness is constant all the way around the bearing (or in the bits we haven't cut out). In fact, the shaft would not adopt this position unless there was zero load on it. But fluid-film bearings are not effective under such conditions – they need to have some load to support. Let's see why this is.

In 1883, at about the same time as Petrov published his equation, the memorably named engineer, Beauchamp Tower, was investigating railway axle bearings for the Institution of Mechanical Engineers. As part of his researches, he was trying to measure what he thought would be the relatively low pressures in the oil film. When the cork in the end of his pressure gauge tube kept blowing out he realized he was on to something. Clearly the oil pressure was far higher than expected. Soon afterwards, Osborne Reynolds, the first Professor of Engineering at Manchester, investigated the theory behind this phenomenon and the science of what came to be called 'hydrodynamic lubrication' was up and away. **▼Some hydrodynamic magic▲** explores a bit of what came out of Tower's observations.

▼Some hydrodynamic magic▲

Imagine looking end-on at the bearing arrangement shown in Figure 3.24 in the previous section. What you would see is sketched in Figure 3.25(a). If a load is now applied to the journal, you would find it has moved slightly from the concentric location to the off-centre one sketched in Figure 3.25(b). The journal is still fully free to rotate and is fully supported by the oil film, but with the rotation axis in a new position. If you were to draw a line between the two axis positions and extend it both ways, you'd find that it bisected the oil film at the regions of maximum and minimum film thickness. Under load, the journal has adopted an equilibrium position at an eccentricity, e, from the true (concentric) bearing centre and attitude angle, ϕ, in relation to the line of action of the load.

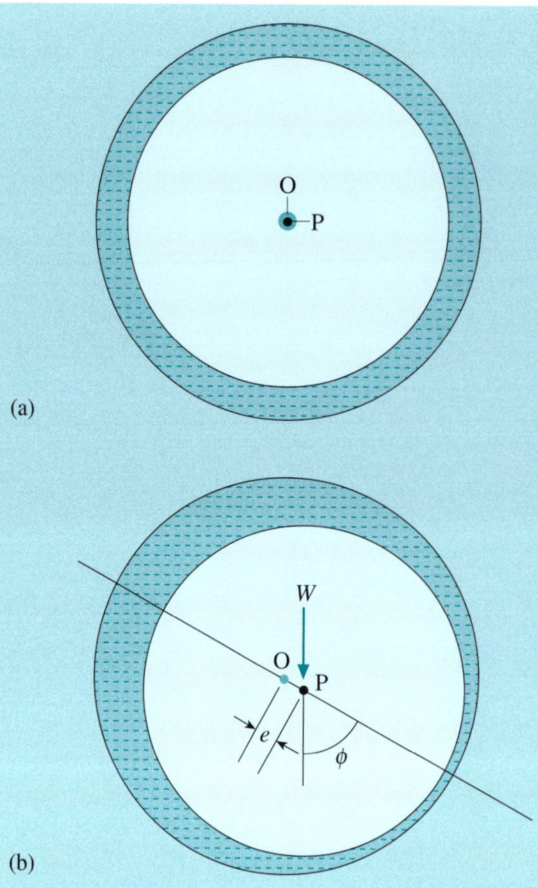

Figure 3.25 End-on view of a hydrodynamic journal bearing (a) under zero load and (b) loaded

You should bear in mind that the eccentricity shown is greatly exaggerated in Figure 3.25, as is the oil film thickness itself. In Figure 3.25(a), the concentric, uniformly shaped oil film has a thickness equal to the radial clearance which, for a bearing of the size depicted, would be about 0.016 mm, barely visible. A journal eccentricity which is a small portion of this might be about 0.003 mm for a reasonable load. So, if the oil film thickness is so small and journal eccentricity even smaller, what is the big deal about such small changes?

As it happens, they are enormously significant. You can see from Figure 3.25(b) that, in a loaded journal bearing, the oil film shape converges to a minimum then diverges to a maximum. The oil is still separating the journal from the bearing housing, but it's being dragged into a converging gap. But, as a liquid, lubricant oil is pretty incompressible. So what happens to the oil in the narrowing gap? A region of very high pressures develops all by itself in the converging section. This resists the oil entering the narrowing gap (and boosts the oil leaving it). Pressures multiplied by the area upon which they act create a net force – in this case, a force large enough to balance the applied load that caused it to develop in the first place. The more load, the more eccentricity, the more convergence, the higher the pressures; hey presto – the load is still supported!

This pressure field is not dependent on any supply-circuit pumps and pressures. These are just to get the oil into the bearing so it can do its stuff. The shape of a pressure-field profile is quite complex and resembles a hump or hill. In the circumferential direction it will look something like that in Figure 3.26.

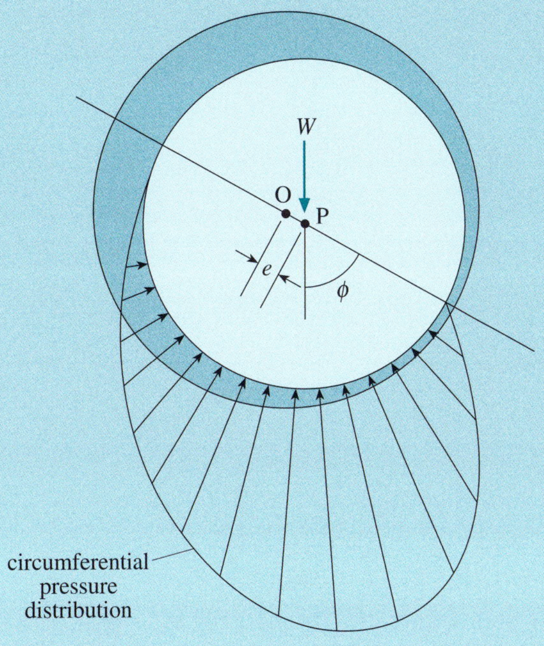

Figure 3.26 Circumferential pressures in a hydrodynamic journal bearing (shaft rotating anti-clockwise)

The oil doesn't stay in one place though; there is a flow in the circumferential direction and some of the oil continues to be dragged right round the bearing. There is also a flow axially (i.e. sideways) as the oil escapes through the clearance space, boosted by the pressure field. So in two dimensions the pressure field looks a bit like that in Figure 3.27, imagining the oil film to be unwrapped and flattened out.

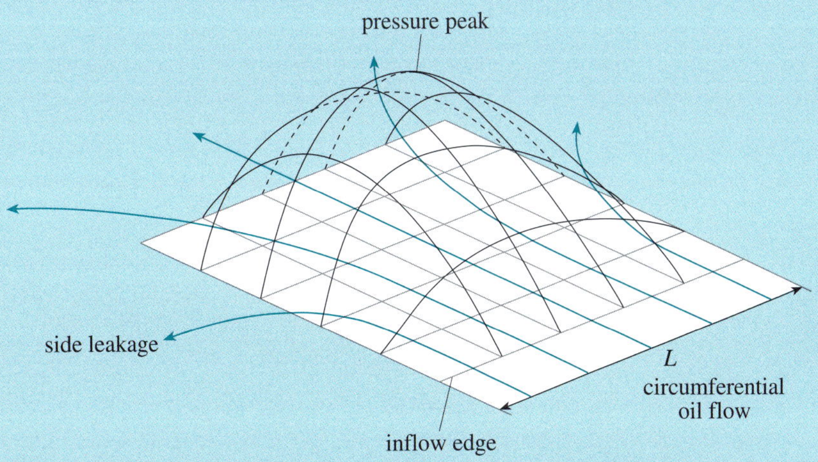

Figure 3.27 Circumferential and axial pressure distributions in a hydrodynamic journal bearing

Figure 3.28 shows another pressure field developed from a computer simulation model of a bearing, created for an Open University research project. There is something quite magical to me about hydrodynamic bearings – whatever the applied load, they will automatically provide their own pressure field to support it. Up to a point of course.

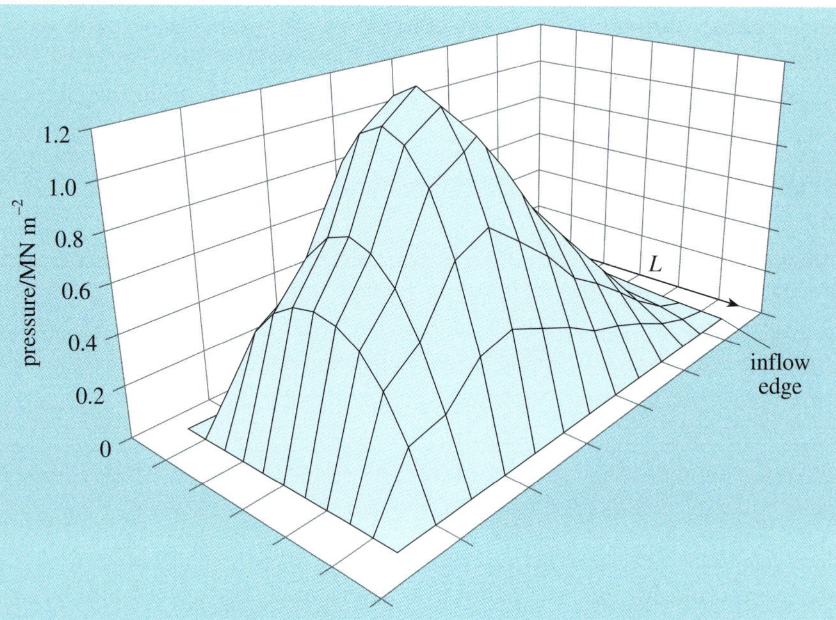

Figure 3.28 Computer model of pressure field in a hydrodynamic journal bearing

In principle, only two things are needed for a hydrodynamic fluid-film bearing to work: well, three really. One is that there must be a convergence in the shape of the lubricant film. The second is that there should be a difference between the velocities of the journal and bearing surfaces. Usually one surface is stationary and the other is moving. In the Dinorwig case (and many others), the journal rotates within a stationary bearing housing.

SAQ 3.15 (Outcome 3.8)

What is meant by hydrodynamic lubrication and what conditions are necessary for it to be invoked?

There are some practical issues to be addressed for bearings operating in the real world, when full-film lubrication may not always be guaranteed. Not the least of these is the choice of bearing materials. The vast majority of journal bearings in industry comprise a rotating steel journal supported in a bearing housing. So what material should the bearing surface itself be?

9.2 Problems with hydrodynamic bearings

The first problem with hydrodynamic bearings is, of course, that they only work when the pressure field has developed – when they've got going, in other words. Once going and having developed full-film lubrication, they can run indefinitely. If though the load gets too large, the eccentricity will approach the limit available within the clearance space. The oil film pressures will then get very high. There's nothing wrong with that – oils can sustain huge pressures – but if you look back at Petrov's equation, it predicts that as the oil film gets thinner (C_r decreases) the power absorbed increases. Although Petrov's equation is for concentric operation, this is indeed what happens. The power absorbed increases and turns into heat, so that locally the oil gets very hot, perhaps too hot to sustain the pressure field (because the viscosity drops off), bringing the risk of surfaces contacting and leading to massive localized friction and heat, and consequent bearing failure.

However, the local decrease in viscosity due to the temperature increase is at, or very near, the region where the pressures are high. Pressures increase viscosity, though not by as much as is lost by the temperature increase. Still, every bit helps.

Overall, it's been found best to design bearings to operate with an *eccentricity ratio*, i.e. e/C_r, of no more than around 0.6 or 0.7. You need some eccentricity, otherwise the bearing wouldn't work. If there is no load, no eccentricity, no pressures, there is no load support reaction – effectively, the journal is not supported. Remember that hydrodynamic bearings aren't effective under zero load conditions? Well, this is why. With zero load, the journal may become unstable and vibrate about the bearing centre. For example, if there's a small, repeating, disturbing force due to something being slightly out of balance, the rotating journal centre may not stay in one place, but may itself rapidly move around in an orbital path.

Figure 3.29(a) overleaf shows just such an orbital path recorded during tests on a marine bearing. Figure 3.29(b) is a picture of the test bearing being installed on a pinion shaft. The journal or shaft diameter at the bearing was approximately 190 mm, and the radial clearance – the oil space thickness with the shaft concentric – was about 0.2 mm. The pinion itself mated with a large gear. During the tests, a large journal orbit was induced by arranging for the load carried by the bearing to be very small, a situation most unlikely in a real ship, but possible nevertheless. Proximity inductive transducers monitored the position of the journal as it operated at steady speed. Figure 3.29(a) shows what happened when the load was significantly reduced. This is a trace of the dynamic orbit of the journal, as it whirls a fraction of a millimetre out of true. The orbit diameter was up to about 80% of the radial clearance, i.e. the available space for the journal to move in. Instead of the journal centre of rotation adopting and settling at an equilibrium position, the centre of rotation itself rotated in an orbit, probably excited by the gear teeth meshing. In other words, the journal was wobbling as it rotated. In a real ship, this would affect the gear teeth alignment and greatly add to the ship's noise signature. Figure 3.29(c) shows how the use of an adjustable bearing could suppress such an orbit. For the same load and running conditions as for Figure 3.29(a), the journal orbit was reduced to one-twentieth of its original size.

Interestingly, the Dinorwig turbine bearing seemed to be suffering similar problems when it was run at reduced loads, a situation it wasn't originally intended for. The bearing loading was less than usual and, importantly, additional forces resulted from vortices or discontinuities due to the less-than-optimum water flow through the turbine blades. These were manifested as a notable shaft vibration at the lower end of the plant, signifying the onset of a stability problem when the load on the bearing decreased as the power output of the generator was reduced. Running the plant in this condition for any length of time is not a desirable situation. The vibration may lead to bearing surface-to-surface contacts, bearing damage, fatigue problems in the shaft, problems with seals and clearances, etc. Better not to have the vibration. The problem is, the market for Dinorwig's power has changed and it is to their benefit to sell some reduced-output power for longer periods, rather than the flat-out power available in stand-by mode as originally intended. In other words, a change in operating circumstances has provoked a new set of problems in a hitherto perfectly reliable plant. This happens a

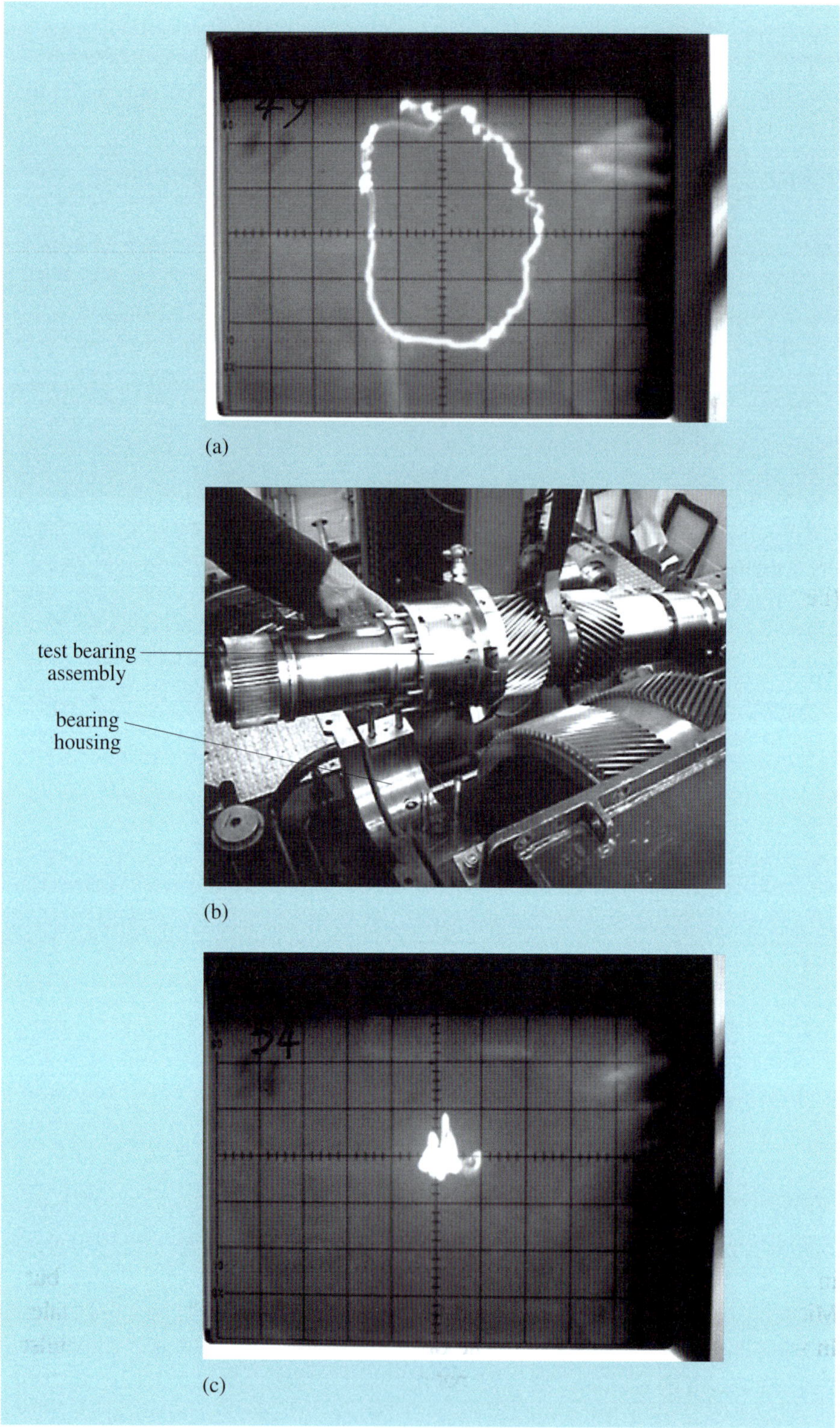

Figure 3.29 Tests on a marine hydrodynamic journal bearing (b) that map the whirling of the shaft in the bearing (a) over a fraction of a millimetre or (c) with much improved stability. In (a) and (c) the images represent the axial position of the shaft, with each major grid line corresponding to about 0.025 mm

lot in engineering – it may suit the accountants to change the design parameters after the event, but it often gives engineers grey hairs!

9.3 Tilting-pad bearings and others

There have been a number of ideas and developments over the years to address the problem of instability in lightly loaded hydrodynamic bearings. One of the most significant is the tilting-pad bearing, also known as the pivoted-shoe bearing. The principle was invented early in the twentieth century, quite independently by two engineers at about the same time, each working without knowledge of the other (no internet in those days!). As it happens, they were working on thrust bearings, but the idea was subsequently adapted to journal bearings.

The basis of the new design was that instead of having a full circular bearing, it had a number of bearing pads, each supported by a pivot such that the pad was free to tilt. Each pad would then be able to adopt a hydrodynamic converging wedge shape on its own to suit the loading condition on the pad. Figure 3.30 shows the idea in principle. In this case there are four pads. Each pad's pivot is central, which is the best place for a bearing which has to cope with a journal that may rotate in either direction (i.e. forward and reverse).

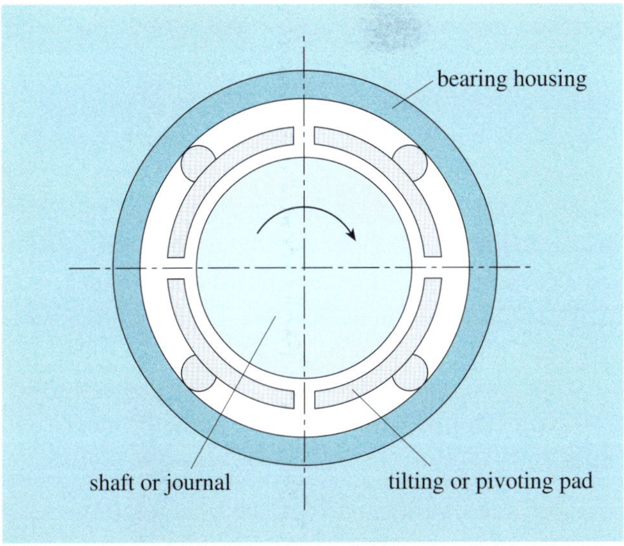

Figure 3.30 Schematic of tilting-pad bearing

The two engineers were Alfred Kingsbury in the USA and Anthony Michell in Australia. Kingsbury was the first to make and demonstrate a device, but Michell was the first to publish and take out a patent. Thereby hangs a tale. In the USA, priority is given to the originator in chronological terms, whilst elsewhere it is the first one to the patent office. Thus, Michell got the patents in the UK and Australia, but not in the USA, whilst Kingsbury got the patent in the USA, but not elsewhere. Not that there was any animosity between the two men: both were to pay generous tribute to the work of the other. But even now the types of bearing are referred to as the Kingsbury pivoted-shoe bearing in the USA, and the Michell tilting-pad bearing in the UK. Early applications were for hydraulic machinery (Michell worked on hydroelectric schemes in Australia), but they were ideally suited to propeller-shaft bearings in ships. They were so effective in reducing friction that it has been estimated that the Royal Navy saved £500 000 in 1918 on coal costs alone – perhaps worth twenty times that in today's money – by using them. Their clear advantages led Michell Bearings Ltd to be formed on the banks of the Tyne in Newcastle in 1921. The company is still there, busily engaged in making and refurbishing large bearings of all sorts, particularly of course tilting-pad types.

The two engineers' projects followed a classic pattern of good engineering ideas:

1 They were prompted by a need, in the first instance by problems with hydraulic machinery bearings.

2 Once demonstrated, a whole new market opened up, in this case in marine applications.

3 They were the direct result of close study and engineering application of academic theory, in this case Reynolds' theory of hydrodynamic lubrication (ironically prompted by an engineer's experiments and observations in the first place).

4 They are (deceptively) simple and obvious – with hindsight!

As you might expect, a number of other ideas have been developed to improve the stability and accuracy of hydrodynamic journal bearings, and many patents have been granted. Figure 3.31 shows some of the more successful designs. In principle, these show ingenious ways to create the hydrodynamic converging wedge shape for the oil film without relying totally on the need for the journal to adopt an eccentric rotational centre position. By careful machining of the bearing housing, two, three, four or more converging wedge shapes are built in to the bearing. Those shown on the left side of Figure 3.31 are designed for a journal rotation in both directions, those shown on the right are for a single direction only. Note that the 'offset halves' bearing is actually a concentric bearing arrangement, but with two halves adjusted in position relative to each other to form two converging wedge shapes. All of these bearings can offer improvements, but at the cost of machining and assembly complexity. Bear in mind that the shapes shown are greatly exaggerated and in reality the wedge shapes would not be noticeable by eye. As you can imagine, machining some of them accurately is very challenging.

The Dinorwig turbine and pump support bearing is of the tilting-pad type. The journal diameter at the bearing position is 1210.0 mm. There are eight bearing pads, each pivoted at the rear surface centre position to allow for both directions of rotation. The length of bearing surface on each pad in the shaft circumferential direction is 390.0 mm, the width in the shaft axial direction is 350.0 mm. Another important difference is that the oil supply is not pumped directly to the pads. The bearing is arranged so as to run semi-submerged in oil. When running, the oil level is about 20% up the surface of the pads. This is not the most efficient form of oil supply but is at least reliable – there's always plenty of oil around each pad. Let's have a go at some power-loss calculations. First we can have a graphic reminder of the importance of viscosity. The main data for the bearing, supplied by the then GE Energy (UK) Ltd, states that the oil used is VG 68 and operates at a working temperature of 65 °C, for which the absolute viscosity is 0.0195 N s m^{-2}.

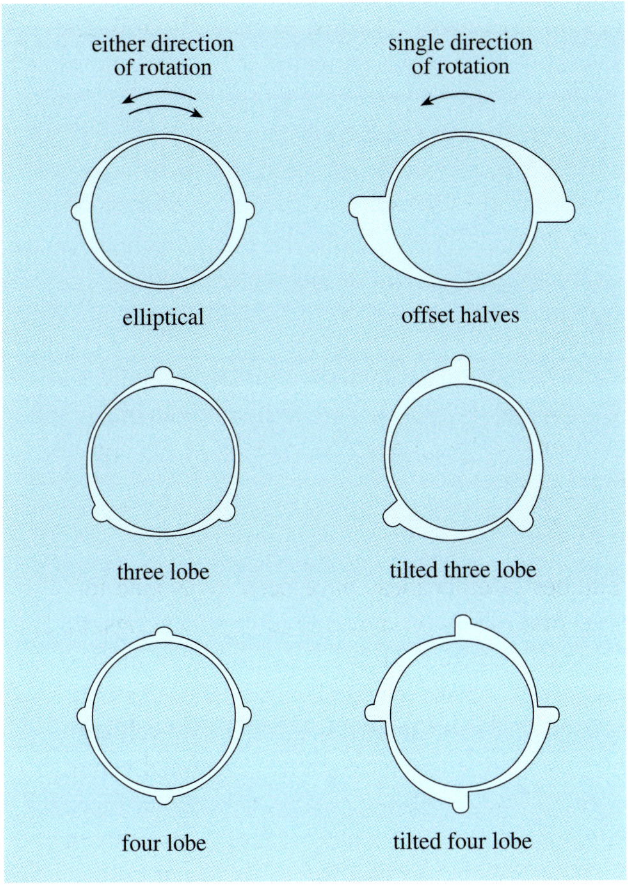

Figure 3.31 Schematics of different fixed-arc bearings

Now it's time for you to attempt some SAQs.

SAQ 3.16 (Outcome 3.7)

Assuming that the data for the bearing are still as specified for SAQ 3.13 (page 178), and that the bearing is still a full circular bearing (no pads yet), determine the percentage reduction in power absorbed using the known viscosity at the working temperature compared with the maximum limit at 40 °C.

SAQ 3.16 shows how important it is to know the actual viscosity values at which the bearing is operating, or to design it such that it operates at the required temperature. If, for example, the temperature was too low there might not be enough power available to drive the shaft at the correct speed.

Now take into account the effect of the pads, or more correctly the gaps between the pads. With reference to SAQs 3.12 and 3.13 (pages 177 and 178), we can quickly determine the effect of the revised viscosity and the gaps on the power absorbed. Perhaps you can have a go at this in SAQ 3.17. You should be able to do this without having to work out the full Petrov equation.

SAQ 3.17 (Outcome 3.6)

Estimate the power absorbed, taking account of the revised viscosity and then the effect of the gaps between the bearing pads.

We've now got an estimate for the power absorbed as about 80 kW. The figure estimated by GE Energy (UK) Ltd was actually 142.4 kW. Our model was based on Petrov's equation, which is stretching things a bit. The Petrov equation is for a full circular, plain bearing running with perfect concentricity. The GE Energy model has been developed to take account of the real situation with the tilting-pad bearing. We still also have some doubts about the real temperatures and running clearances. Still, the Petrov estimate is not too bad for a 'ballpark' estimate, and at least points towards more confidence in the GE Energy model. (If the two were different by a factor of 10, say, I'd be more concerned.) It's always worth checking computed results with the simplest reliable model to hand, particularly if an actual computer model is used.

10 LEARNING OUTCOMES

After studying this part of Block 4 you should be able to do the following.

3.1 Distinguish between friction and wear, and use coefficients of friction and wear in estimating their effects. (SAQs 3.1 and 3.2)

3.2 Carry out engineering calculations of mass, second moment of mass, kinetic and potential energies, as related to a hydroelectric power station generation plant. (SAQs 3.9, 3.10 and 3.11)

3.3 Demonstrate a grasp of the principal properties of fluids, in particular viscosity. (SAQs 3.4 and 3.5)

3.4 Say what is meant by a Newtonian fluid. (SAQ 3.6)

3.5 Interpret currently used units and grading systems for viscosity of oils. (SAQ 3.7)

3.6 Describe the history and development of boundary and full-film lubrication. (SAQ 3.3)

3.7 Assess the importance of the various factors contributing to the energy loss and power absorbed in various types of full-film bearing. (SAQs 3.12, 3.13, 3.14, 3.16 and 3.17)

3.8 Define 'hydrodynamic lubrication' and outline the conditions necessary for it to exist. (SAQ 3.15)

3.9 Describe the practical requirements of, and issues related to, the specification of full-film bearing alloys. (SAQ 3.8)

ANSWERS TO EXERCISES

Exercise 3.1

The other functions of the bearing are to allow as free rotation as possible of the wheel, to transmit the weight of the cart to the wheel, to translate the forward rolling motion of the wheel to the cart, and to resist sideways forces on the wheel.

The phenomenon, as you should recognize, is friction.

Exercise 3.2

PTFE (Teflon®), the 'non-stick' material used in many applications, would seem to be a good choice. A highly polished, hard surface of stainless steel, for example, might be effective. It must depend on what material the other bit is made from. Presumably not all materials can be used inside the body owing to problems of toxicity or corrosion.

Exercise 3.3

Here's the list I came up with:

low shear resistance
protects parts from corrosion
wide operating temperature range
good heat capacity
inert to typical seal materials
long lasting
miscible with modifying additives
non-toxic
plentiful and (relatively) cheap.

How did you get on? Did your list look anything like mine?

Exercise 3.4

γ is dimensionless as it is a ratio of two lengths, so G must have the same units as τ, i.e. N m^{-2} or Pa.

Exercise 3.5

τ is shear stress in N m^{-2}, or Pa.

$$\frac{\mathrm{d}\gamma}{\mathrm{d}t}$$

the rate of shear strain, is in s^{-1}.

So from Equation (3.5):

$$\tau = \eta \frac{\mathrm{d}\gamma}{\mathrm{d}t}$$

The SI units of η are:

$$\frac{\mathrm{N\ m^{-2}}}{\mathrm{s^{-1}}} = \mathrm{N\ m^{-2}\ s}, \text{ or Pa s}$$

(i.e. pascal seconds).

Alternatively, τ is shear stress in N m^{-2}, or Pa:

$$\frac{du}{dy}$$

the velocity gradient, is in:

$$\frac{\text{m s}^{-1}}{\text{m}} = \text{s}^{-1}$$

So the SI units of η are:

$$\frac{\text{N m}^{-2}}{\text{s}^{-1}} = \text{N m}^{-2}\text{ s, or Pa s}$$

(i.e. pascal seconds).

Exercise 3.6

In terms of units:

$$v = \frac{\eta \text{ N m}^{-2}\text{ s}}{\rho \text{ kg m}^{-3}}$$

Since 1 kg = N m^{-1} s^2, this simplifies to dimensions of m^2 s^{-1}.

ANSWERS TO SELF-ASSESSMENT QUESTIONS

SAQ 3.1

(a) From Table 3.1, $\mu_s = 0.6$ under these conditions, and the horizontal force required is given by Equation (3.1):

$$\begin{aligned} F_{\text{lim}} &= \mu_s R \\ &= \mu_s mg \\ &= 0.6 \times 5.0 \times 9.8 \text{ N} \\ &= 29 \text{ N} \end{aligned}$$

So the limiting frictional force is 29 N.

(b) The situation is similar to that depicted in Figure 3.8(b), but with the frictional force acting uphill in opposition to the $mg \sin \theta$ component of the skier's weight – the frictional force acts in opposition to the likely motion. At the moment of motion, the force acting uphill is the limiting friction, so from Equation (3.1):

$$\begin{aligned} F_{\text{lim}} &= \mu_s R \\ &= \mu_s mg \cos\theta \end{aligned}$$

This must equal the component of the skier's weight acting down the slope, so that:

$$\mu_s mg \cos\theta = mg \sin\theta$$

giving:

$$\begin{aligned} \mu_s &= \frac{\sin\theta}{\cos\theta} \\ &= \tan\theta \end{aligned}$$

From Table 3.1, $\mu_s = 0.05$.

$$\begin{aligned} \theta &= \tan^{-1} 0.05 \\ &= 2.9° \end{aligned}$$

So the slope must be at an angle of 2.9°.

Note that this gives you another simple method of determining μ_s between an object and a surface. Just increase the slope until the object moves and then measure the angle. Its tangent equals μ_s.

(c) First you must calculate the retarding force, F, on the car. You know from Newton's Second Law that $F = ma$, and you also know that the final speed $v = 0$ at time $t = 7$ s. The initial velocity is:

$$u = \frac{50 \times 10^3}{3600} \text{ m s}^{-1} = 13.9 \text{ m s}^{-1}$$

Thus the equation for a is:

$$a = \frac{v - u}{t} = \frac{-u}{t} \text{ (deceleration)}$$

So the magnitude of the retarding force on the car is:

$$F = \frac{mu}{t}$$

If the force between each lining and drum is F_B, this is the equivalent to the normal reaction force R in, for example, Figure 3.4(a). So the frictional retarding force from each brake is $\mu_d F_B$ and there are four of them. Hence:

$$F_B = \frac{mu}{4\mu_d t} = \frac{800\times 13.9}{4\times 0.3\times 7} = 1323 \text{ N}$$

So the force between each drum and its lining is about 1.3 kN.

SAQ 3.2

You may have chosen one each from these lists, or something similar.

Sliding contacts where wear is undesirable:

- pistons sliding into their housings;
- objects being slid from one place to another (e.g. pushing a refrigerator into position under a kitchen unit, a bed being pulled for sheet changing, a pan being moved across a hotplate, an oven shelf being relocated, a spade or shovel being pushed into sand, gravel, soil, etc.);
- meshing gear teeth and chains engaging cogs.

Sliding contacts where friction is desirable:

- objects left otherwise unrestrained on sloped surfaces (e.g. crockery in transit on a tray, builders working on pitched roofs, snow lying on pitched roofs;
- ropes wound round capstans;
- any objects on rising or descending conveyor belts;
- brake drums, discs, etc.

SAQ 3.3

Boundary lubrication occurs when two surfaces are in sliding contact but the friction is reduced by the presence of a lubricant. This has the effect of reducing the level and numbers of enmeshed surface asperities, thereby reducing in turn the localized deformations and welding and hence reducing resistance to relative movement. With full-film lubrication, the two surfaces are completely separated by a lubricant film and the resistance to relative movement is due only to that of the lubricant itself, (actually its viscosity, as discussed later).

Examples of boundary lubrication are many and varied. For example the oiling of a creaking door or gate hinge, waxing skis or the lubrication of a piston in an engine cylinder come to mind. Full-film lubrication is found in the main bearings of an engine crankshaft, in vehicle gearboxes and in bearings in industrial machinery, amongst others.

SAQ 3.4

(a)

$$\frac{\rho_{\text{oil}}}{\rho_{\text{water}}} = 0.891$$

so $\rho_{\text{oil}} = 0.891\times 1000 = 891 \text{ kg m}^{-3}$

(b) $\text{mass} = \rho_{\text{oil}} \times \text{volume} = 891\times 0.001 = 0.891 \text{ kg}$

(c) weight = mass × g = 0.891 × 9.81 = 8.74 N

(d) specific weight = weight ÷ volume =

$$\frac{8.74}{0.001} = 8740\ \text{N m}^{-3}$$

SAQ 3.5

The lower-density substance will accumulate on the top (olive oil in this case). Specific gravity is just density relative to water, so lower density also implies lower specific gravity.

SAQ 3.6

$$\rho_{\text{oil}} = 891\ \text{kg m}^{-3}$$

$$\nu = 32\ \text{cSt} = 0.32\ \text{St} = 0.32\ \text{cm}^2\ \text{s}^{-1} = 32.0 \times 10^{-6}\ \text{m}^2\ \text{s}^{-1}\ \text{in SI units}$$

From Equation (3.6):

$$\eta = \nu\rho = 32.0 \times 10^{-6} \times 891\ \text{Pa s} = 0.0285\ \text{Pa s}$$

So the coefficient of dynamic viscosity for the oil is 0.0285 Pa s.

SAQ 3.7

From Equation (3.6):

$$\begin{aligned} \nu = \frac{\eta}{\rho} &= \frac{0.0891}{891}\ \text{m}^2\ \text{s}^{-1} \\ &= 100 \times 10^{-6}\ \text{m}^2\ \text{s}^{-1} \\ &= 1\ \text{St, i.e. } 100\ \text{cSt} \end{aligned}$$

Thus, the oil would be VG 100.

SAQ 3.8

White metals are (usually) tin-rich alloys, whose main components are tin, antimony and copper with a range of other additives chosen empirically to confer the following attributes:

- resistance to seizing as material scores off onto the shaft;
- strength to withstand compressive loads;
- deformability sufficient to be shaped for function;
- strength in the face of cyclic loading;
- resistance to particulates becoming embedded;
- resistance to corrosion;
- structure of hard particles in a soft matrix or vice versa.

SAQ 3.9

The rotor has substantially more mass at markedly larger radii than the shaft, so this is expected to dominate rotational inertia.

The second moment of mass of a hollow rod/disc is given by Equation (3.9):

$$I = \tfrac{1}{2} m \left(R^2 + r^2 \right)$$

For the rotor, the outer radius $R = 4.554/2$ m and the inner radius $r = 1.000/2$ m (where the shaft has been removed) – notice the factor of 2 that divides the given diameters. This gives a value of I for the rotor of:

$$\tfrac{1}{2}(374\times10^3\times(2.277^2+0.500^2)) = 1.02\times10^6 \text{ kg m}^2$$

Taking equivalent data for the shaft:

$$\tfrac{1}{2}(51\times10^3\times(0.500^2+0.100^2)) = 6.63\times10^3 \text{ kg m}^2$$

As expected, the rotor dominates.

SAQ 3.10

From Equation (3.7):

$$\begin{aligned}\text{kinetic energy } &= \tfrac{1}{2}I\omega^2 \\ &= \tfrac{1}{2}\times(1.0773\times10^6 \text{ kg m}^3)\times\left(\frac{500\times2\times\pi}{60}\text{ rad s}^{-1}\right)^2 \\ &= 1.48\times10^9 \text{ J}\end{aligned}$$

So the kinetic energy of the rotating assembly is about 1.5 GJ.

SAQ 3.11

Assume that all the kinetic energy of the rotor assembly can be transferred to the car with no losses, the rotor assembly stops and the car contains all the energy in potential form.

Equate potential energy (mgh) of the car to the kinetic energy in the rotor (1.5 GJ). Take $g = 10$ m s^{-2} and $m = 1.5\times10^3$ kg.

$$mgh = 1.5\times10^9 \text{ J}$$

$$h = \frac{1.5\times10^9}{1.5\times10^3\times10} = 10^5 \text{ m}$$

So the car would be raised to a height of 100 km (almost 70 miles) – much to the alarm of the occupants, no doubt!

SAQ 3.12

η: viscosity – direct linear influence; more viscosity – more power absorbed

r: shaft radius – direct cubic effect; more radius – more more more power absorbed

L: bearing length – direct linear influence; more length – more power absorbed

ω: shaft speed – direct squared effect; more speed – more more power absorbed

C_r: radial clearance – inverse direct effect; more clearance – less power absorbed.

SAQ 3.13

From Section 5.3, kinematic viscosity at 40 °C = 68 cSt, with a tolerance of ±10%. The highest possible value is therefore 68 + 6.8 = 74.8 cSt. The smallest radial clearance is 0.3 mm. Both these limits, the worst cases, will lead to the maximum torque (see SAQ 3.12).

Kinematic viscosity, ν, is 74.8×10^{-6} m^2 s^{-1}.

Dynamic viscosity, η, is therefore:

$$74.8\times 10^{-6}\times 900 \text{ Pa s} = 0.067\,32 \text{ Pa s}$$

Radial clearance, C_r, is 0.3×10^{-3} m.

Angular speed, ω, is:

$$\frac{500\times 2\pi}{60} = 52.4 \text{ rad s}^{-1}$$

Bearing length, L, is 0.350 m.

Journal radius, r, is 1.210/2 m = 0.605 m.

All the data is now in the correct units ready for putting into Petrov's equation (Equation 3.13). So:

$$\Gamma = \frac{2\pi\times 0.06732\times 0.605^3\times 0.35\times 52.4}{0.3\times 10^{-3}} \text{ N m} = 5726 \text{ N m}$$

$$\text{power absorbed} = 5726\times 52.4 \text{ W} = 300 \text{ kW}$$

$$\text{total energy lost} = 300\times 10^3\times 60\times 60 \text{ J} = 1.08 \text{ GJ}$$

So the total energy lost after one hour's running is 1.08 GJ.

SAQ 3.14

The key is in Step 2, in the line:

'(this assumes the fluid film extends right round the bearing)'

What would happen if it didn't? For example, what if the bearing area wasn't $2\pi rL$ but, say, $1.6\pi rL$?

Then the new power absorbed will be less, in the ratio 1.6 to 2, i.e. 80% – in other words a 20% reduction!

SAQ 3.15

Hydrodynamic lubrication is a self-generating and self-sustaining form of full-film lubrication, whereby the two surfaces are separated by an oil film in which pressure fields are generated automatically to balance any applied load. Necessary conditions are a relative velocity between the two surfaces, a converging oil film shape and a lubricant of sufficient viscosity.

SAQ 3.16

VG 68 at 40 °C has a mean kinematic viscosity of 68 cSt or, in SI units, 68×10^{-6} m^2 s^{-1}. The dynamic viscosity is given by Equation (3.6):

$$\begin{aligned}\eta &= \nu\rho \\ &= 68\times10^{-6}\times900\ \text{N s m}^{-2} \\ &= 0.0612\ \text{N s m}^{-2}\end{aligned}$$

At 65 °C, the dynamic viscosity is known to be 0.0195 N s m^{-2}.

From Petrov's equation (Equation (3.13)), $P \propto \eta$.

The percentage drop is therefore given by:

$$\left(\frac{0.0195-0.0612}{0.0612}\right)\times100=-68\%$$

So the power absorbed in the bearing drops by 68%.

SAQ 3.17

From SAQ 3.16, the power will now be:

$$\left(\frac{100-68}{100}\right)\times300=96.0\ \text{kW}$$

For the tilting-pad bearing, there will be a further reduction due to the gaps between pads. Each pad length is 0.39 m and there are eight of them. The wetted pad length is thus 8×0.39 m $=3.12$ m. The full bearing wetted length is $\pi D=\pi\times1.210$ m $=3.801$ m.

From Petrov's equation, $P\propto$ area of bearing.

Therefore the revised power loss will be:

$$\frac{3.12}{3.801}\times96.0=78.8\ \text{kW}$$

The power absorbed using the viscosity of the oil at 65 °C and tilting-pad bearings is 78.8 kW.

ACKNOWLEDGEMENTS

Grateful acknowledgement is made to the following sources for permission to reproduce material within Part 3 of this block.

Figures and cover

Figures 3.2 and 3.3: Reproduced by permission of First Hydro Company; *Figure 3.10:* Reprinted from *Theory and Clinical Applications,* Steven Kurtz, 2004, with permission from Elsevier; *Figure 3.11:* AFM profile supplied by Dr Sharon Moore, The Open University; *Figure 3.12:* Courtesy of Dr. T. Ellis, Brunel University; *Figure 3.14:* RHP Bearings Ltd; *Figure 3.15:* © Bettmann/CORBIS; *Figure 3.29(b):* Lubrication Research group, The Open University.

Course team acknowledgements

Part 3 of this block was prepared for the course team by Keith Martin and George Weidmann.

Trademarks